Living it up in France

HETTIE ASHWIN

Published by Slipperygrip 2020

Living it up in France.
Copyright Hettie Ashwin

All rights reserved

The moral right of the author has been asserted.

This book is sold subject to the condition that it shall not, by way of trade or otherwise, be lent, re-sold, hired out, or otherwise circulated without the author's proper consent in any form other than that in which it is published and without a similar condition including this condition being imposed on the subsequent purchaser.

PAPERBACK
ISBN: 9782491490041
POCKET EDITION
ISBN: 9782491490003

www.facebook.com/alacrity.vivacity
www.hettieashwin.blogspot.com

Books by Hettie Ashwin

<u>Humour</u>
Literary Licence
The Reluctant Messiah
Mr Tripp buys a lifestyle
Barney's Test
The Truffle War
Fat Bits
Boat to Baguette
Murder! Mayhem! and lesser cuts of meat.
I'd rather glue me nut sack to a bullet train

<u>Thriller</u>
The Crowing of the Beast

<u>Speculative fiction</u>
The Mask of Deceit
Pi - trilogy. Bk 1

<u>Short Stories</u>
After the Rains & other Stories
A shilling on the Bar

<u>Novellas</u>
A Strange kind of Paradise (series)

"If it's not fun, you're not doing it right."
—
Bob Basso

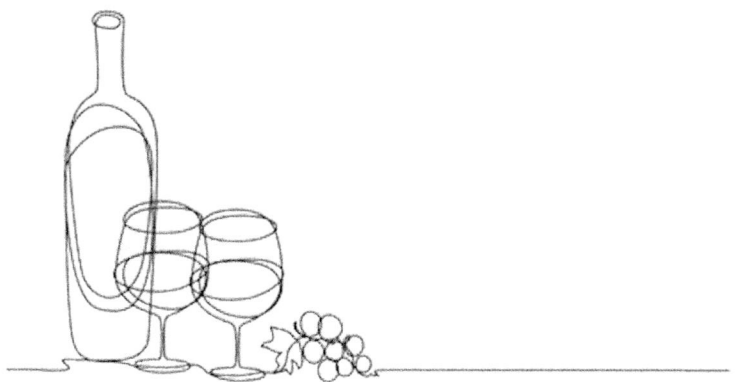

Our home in Cote d'Armor, Brittany.

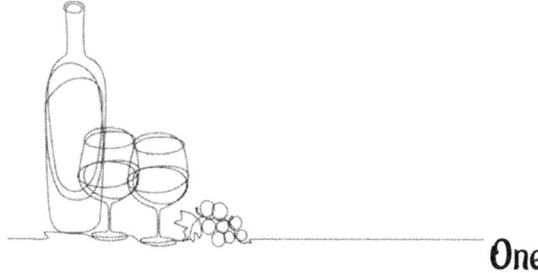

One

And we were at it again, scheming our way into travel, renovations, and a love of good wine.

Not that we needed any coaxing in regard to wine, but after renovating a boat for more years than I care to remember, I was a bit over the painting, the sanding and the visits to the hardware store.

We thought we were so clever. Two Australians who decided to buy a house in France. We sat back in our little house in Brittany, the paperwork completed, the bank happy, our vocabulary expanding and contemplated all the things we were going to do. Our *to do* list grew, and grew, and grew. Three lifetimes might not be enough, and we only had one, and that was getting to its sell by date.

What we didn't want to do was easy.
'I don't want to do much to the house, just small renovations.' I said it as I tried to shut the old window in the kitchen and used the pull-up, push-down, give it a shove technique.

'I'm not keen on a lot of gardening,' Boomie said as he scraped the mud from his shoes.

'We should go somewhere.'

It sounded like a fine idea. After all, this is why we came to France in the first place. To go places, see things and just … well, just enjoy life. How hard can it be?

And so, we pulled out our trusty old map. It had been folded, concertinaed, spread out, pawed over, and I could almost hear it sigh as once again we looked for adventure, little post-it notes at the ready.

'But first,' Boomie said, 'we should get ourselves organised. I put the kettle on, brought out the cake and settled in for the afternoon.

'No, I mean make a list of the things we must do, then the things we should do, *then* the list of things we want to do.'

'Oh.' I was already planning what hats to take and how many changes of underwear.

Two

Our *things we must do* was a large one.

It was January. It was cold, and the bunker diesel to keep warm was eating into our fun budget. I found a slug crawling inside to keep warm, such was the depths of winter in Brittany. We were wearing our woollens inside, while we tried to limit our heating bill.

The French like their creature comforts like any normal human being, and two Australians from warmer climes were in need of creature comforts. We went to the big supermarket down the hill and purchased, what we'd known in Australia, a 'kero heater'.

They had had a make-over in the 50 odd years since we'd seen them as kids. Now they were electric start, came with an easy dispense bottle and didn't smell.

We lugged our prize up the hill in our wheelbarrow and with a quick flick of the starter button, thought we would be warm, toasty warm, enough to shed a layer or two.

'It won't start.' I said as I looked at the 'made in China' sticker. Usually, it is a case of sometimes you win, sometimes you lose in the People's Republic of China. Boomie had a go and we tried to fathom the instructions.

These days instructions are non-literary, for the non-literary. Funny little pictures that have no relevance to the object, show you how to pull the machine apart while counting the screws.

'Batteries.'

Naturally, it didn't come with batteries.

'I'll go,' I said hoping he would say no.

The heater sprang to life and we huddled over it remarking how warm it was, how efficient it might be at heating the kitchen and working out how much it would cost. Being married to an accountant, there is always a bottom line.

We are people who get junk mail. Some spurn the catalogues, but we like to look, we like to peruse, we like to scoff, usually incredulously.

'Will you look at the price of that,' can often be heard at the kitchen table. But when you are hunting for heater fuel, it is a worthwhile exercise. The trick, and there is always a trick is to get the stuff home. We still didn't have a car.

There was a plan to only put the heater on after 4 pm. I became a clock watcher.

The thing to do to keep warm is work.

We started our renovations, keeping in mind our vow not to go too far, too fast, and reminding ourselves that it is an old house, it didn't cost much, we don't want to, God forbid, *overcapitalise.*

Our front steps come up to a landing that had been enclosed many years ago. The brickwork was a little askew and the paint job was slewing off like sunburnt skin. I found a trusty scraper and started to do an epidermis peel. There is nothing like getting big sheets of paint off in one go and I had the chip, chip, peel down to a fine art.

'The trouble you see is …' Boomie took a look at my now bare walls. They were weeping. They were wet.

We looked at our windows. They were crying, weeping like a widow at the graveside.

The trouble was our little heater. The condensation was making our house into a bath-house. All the paint started to bubble. All the window frames started to soak up water like a sponge and then we noticed the smell. Sort of like old mushrooms with something dead.

I tenderly poked a window frame with a fork and the whole thing was like soggy soap.

We looked at the culprit and I swear it gave a Gallic shrug.

'Who, me,' it said and then clicked off. Not only was it making our house condensate at an alarming rate, now we were in peril of gassing ourselves. The fail-safe had clicked off to save us, but we didn't hold much hope for its longevity, considering the bargain price and the manufacturing standards in PRC.

What was the point of keeping warm if your house was going underwater and killing you? What was the point if you needed to open a window?

'Well, we could keep the door to Colditz open.' We had christened the little back room, that was half in the ground, Colditz on our very first night.

'It might just be enough to stop the cold front meeting the warm front and the precipitation from the ceiling.' I lived on a boat for many years, I can't help talking meteorologically sometimes.

Colditz is always about four degrees colder than the rest of the house, but as perverse as it seems, it had the most efficient diesel-fuelled radiator in the house. I could dry my woollens overnight in that room.

The door was opened, the front door to the little atrium closed, the outer door left open to dry the walls and we bought a squeegee from the supermarket to try and keep the water from the windows. It might have been easier just to put on our long underwear, our beanies and gloves. Or just go to bed. Now there's a thought.

There was always a niggling idea in the back of our minds that we were going to fix the old house up a bit. To that end, we purchased tiles at a hardware store in Australia which was closing down. We bought rugs for the day we might put them on the floor and we invested in tile laying apparatus. I've never put a tile down in my life, but Boomie assured me he was an old hand at the skill. And it does require some skill.

I did the prep. I picked the tiles. I left him to it.

It might not have been much, but to us, it was a start. Eight tiles and a skirting edge in the atrium. We marvelled at it. We envisaged what it would look like. We are good at envisaging, having done it for as many years as we've been married, and even probably before that.

And then, twenty-two days after coming home from a trip down under, our toilet blocked … again.

We'd had nothing but trouble from our French drains. I'd seen more poo than I care to see, most of it splattered over our back yard as the pipe burped up our waste with sickening regularity.

And now it was at it again.

I called in the plumber our neighbour, Myriam suggested.

'He's a friend,' she said. 'He is good.'

We hoped so. Because the last plumber, although full of good intentions wasn't much chop. It was like coming to a gunfight with a knife. We had a seriously blocked artery when we first moved in, and he brought a water blaster. It was only later we found out why.

So Hervè came with his fresh-faced offsider. We all peered down the manhole where the sewer pipe reaches the road. I was told to flush.

Now, I had told them that it was completely blocked. Nothing was getting through. I knew if I flushed we would just get a huge burp onto the grass and I'd only just buried the last lot as we had strangers, aka plumbers coming around. But what do I know, I'm only a woman.

I tried to explain with burping noises, and grand hand gestures what would be the likely outcome.

'Non.'

'Ok then …'

Hervè stood near the burping outlet. I leaned out of the window and shooshed him away. He wasn't listening and then I flushed.

There were a few swear words I'd not heard before. The burp landed on his trousers.

If I knew how to say, 'I told you so,' in French I would have said it.

Boomie and the fresh-faced offsider waited at the drain and called out,

'R*ien*.' Nothing.

Now Hervè got serious. He had a water blaster too, but a very powerful one. He shoved that nozzle up the pipe with vengeance and cranked the pressure to max.

We all looked down the hole.

And then there was a collective step back and a few more swear word were heard as 22 days of Ashwin unmentionables glooped and gushed out of the pipe.

The young lad covered his nose and mouth with his jumper. I died a thousand deaths as we saw our poo keep pouring out.

I fancy Hervè said something about striking the mother-lode and as with all things that happen in a small village, the neighbour came over for a look.

But, joy oh joy our toilet was working again and now we knew we had 22 days up our sleeve to get something done.

Myriam came over and through some whispering and hand signals, we were told it would be cheaper for cash.

I hopped to it and came back with a plain envelope.
'Done.'

You'd think we'd be on a roll after a small success. But it doesn't take much to distract the Ashwins, and when Myriam said she was going to the beach and would we like to come, we didn't say,

'Oh, we couldn't possibly go, we are renovating.' We jumped at the chance.

Myriam is one of those people that pack for every eventuality. I wondered where we would sit as the dog, the picnic, the blanket, the son, the market stall paraphernalia, the change of clothes, the rubber boots and the rummage sale box of bits all crowded the back of the car.

We set off in high-spirits listening to something Myriam called music, But to the dog, Pim and us, sounded like bagpipes on drugs.

'You know?' she asked as she turned the volume up.

'No.'

'You don't know? It is very famous.'

Perhaps we didn't get out enough.

Three

The Granite Rose Coast in Brittany looks onto that bit of sea where the English Channel and the Atlantic ocean meet. It is a rugged coast with distinctive pink granite boulders cracking off into the sea. It is prime French real estate.

People the world over like going to the beach. The natives of France are no different, although they treat the beach like a health resort. They extol the benefits of fresh air and plenty of it. They hike with walking sticks and backpacks, and they do a lot of picnicking.

As a rule, French people don't walk and eat at the same time. They are not keen on sitting on a park bench and chowing down on a sandwich. They are much more restrained when it comes to eating in the open. There needs to be a waiter, cold water on the table and a menu at the very least. But picnicking, is a different story. They *pique-nique* like they invented it, and ask anyone and they will say they did.

Myriam brought out all sorts of pate, little rounds of toast, biscuits and things in jars. We couldn't just eat on a rock, we had to find a table. We needed to have a table-cloth, and we needed to set the table. I had been advised not to take anything, as bringing something to eat would be an insult to the host. I couldn't help it. It felt wrong to go empty-handed. We brought some tropical fruit that was on sale at the supermarket. The mango went down a treat as we showed them how to cut and eat it.

I wouldn't suggest that Myriam is unworldly, or ignorant, but life in a small village is sometimes all one needs in the way of stimulation. Australia is an exotic location. Things that we might take for granted are so off the radar for someone who lives in a village like ours, that they might as well happen on the moon. Conversely, what we find unusual, they assume happens all over the world.

'You don't have in Australie?' Myriam asks as she collects seaweed to take home to pickle, make soap and put on the garden.

'Non.'

'You don't like?' Myriam said while she eats seaweed, and Guillaume, her son starts to laugh.

'Non.'

We left the gleaners and went for a walk following a well-worn coastal path. I could imagine people walking the path for millennia. The rock formations jut into the sea and over time they break and are separated creating small shallow bays, islands of sentinels and tiny beaches hidden from the storm. Australian coastal rocks are made for fishing. Here, nature was the only companion as we clambered over the outcrops and marvelled at the splendour of the ocean.

Rock-hopping is fun. There is always a different vista, some outcrop that needs climbing, and shells to discover. I stowed away some interesting rocks in my pocket, while Boomie looked the other way. We weren't sure taking rocks was permitted. We didn't know if it was a national park or there were laws against that sort of thing. With a population of 66 million at the last count, if everyone took a rock, there would be nothing left.

Ambling back to the seaweed eaters we rounded the corner and saw a small cottage that had been built right on the foreshore. It was an impossible location, but had weathered the years in a small lee. It was the stuff of 18 century romantic novels, you know, with the heroine, hair flying in the wind, waiting for her loved one ... and that sort of thing. As we were looking and admiring the owner came out and we soon got talking.

With limited French and English, he told us his family had owned the house for many years. It was their retreat, but it was 'orrible to repair. I can imagine every iron fastening, every bolt would rust the minute you looked at it. His wife was inside making lunch, he couldn't stay and talk. Lunch is an institution in France. Everything stops for lunch, usually for 2 hours. We bade farewell as he scuttled inside to his wife's shout that she wouldn't wait another minute. He gave us a shrug and rolled his eyes as if to say, 'women.' Sometimes you don't need words.

Myriam is rightly proud of her heritage, her little bit of France, and our next stop was a drive around the coast road, taking in the real estate, the sea views and we ended at Tréguier, a village with an impressive cathedral.

Myriam went off to visit a friend and left us to wander the small town, a pastime we enjoy. Our first stop was an ice-creamery with a view of St Tugdual Cathedral. We enjoy history and looking at cathedrals is high on our list of things we like.

To find such a large, well-built church in a small town was a surprise. The information plaques were in English, another surprise.

The site had had a church on it since 970 A.D. and all that remained of that time now was a Romanesque style round tower. The gothic cathedral itself was started in 1339. We walked around it looking up and I maintain I saw it first.

There were quite a few gargoyles around the flying arches, but one in particular, caught our eye. It had a smirk and an enormous erect penis.

I wondered if the stonemason didn't get paid and this was his tribute to his employers. Perhaps it was an 'in' joke about the clergy. Who knows. It certainly was a statement.

The inside was as expected. Wonderful stained-glass, vaulted ceilings and chapels dedicated to dead people. What we didn't expect were more 'naughty' carvings. The choir stalls have figures in altogether compromising positions - think - showing the neighbours all your assets, especially where the sun don't shine!

We were in for a treat as we walked around because the local choir were rehearsing for a Bach recital. This wasn't just a group of amateurs, but a full choir of around 100 people. We sat down in a pew with quite a few other people to enjoy the free performance.

I don't care what anyone says, when you hear a choir in a cathedral it gives you goose-bumps and makes the hairs on the back of your neck stand on end. The soaring voices filling the nave and the choir stall is stirring stuff. The organ was built in 1647 and it was easy to think that the peasants on hearing it might feel a little closer to their maker. It's notes reverberated in my lungs. The singers were well-practiced and I could see this sort of thing making you believe God has an ear for music. We left with organ music ringing in our ears.

Another walkabout hobby of ours is visiting graveyards. We enjoy looking at old headstones, fathoming the history of a family, marvelling at the dates of born and died. So many die young, and yet the family name sometimes survives and populates the whole site, until everyone is related to some degree.

Tréguier cemetery is right in the centre of town, surrounded by an old wall which has gateways protected by old iron bars.

We were looking at the World War I graves, reading the names of the battles where the men had met their end when a very small woman with dead flowers in her hand came up to us.

'Are you English?'

'Australien.'

'Oh. Oh.' She smiled. 'These,' she pointed to the graves, are my boys.'

'Your family?'

'No, but I look after them.' She showed us the dead flowers and pointed to the fresh ones.

We stood and looked at the five gravesites.

'He was my father's friend.' The young man had died at 18 years of age.

'You speak very good English,' I said in French.

'Yes. I have many years to learn, many people are my teachers.' She took Boomie's hand, 'come.'

We followed and she took us to her husband's grave. It was well tended and had a ships steering wheel engraved on the headstone.

'Was he a Captain?'

'Yes, you know this?'

We then told her about how we lived on a boat in Australia for many years and it was enough to cement a friendship.

'I am a Captain too.' We looked at this small woman who would be just under 5 foot. 'I cannot stand for very long,' she said. 'I am old.'

'Very old?'

'You must guess,' she said with a wink in Boomie's direction. I swear she was flirting with my husband.

'80?'

'92.'

She had the most marvellous skin for a 92-year-old woman. Australians don't weather very well in comparison.

'Would you like to see my house?'

How could we refuse an offer like that? Whether we look trustworthy, too old to get up to any mischief or just a little gormless, people often take us into their lives with only the merest of introduction.

Madam Bichue took my arm and steadied herself for the walk to her house, just around the corner and up a hill. Boomie took up the rear with her basket in the crook of his arm.

Her house wasn't just a small two-bedroom affair. It was a Georgian style, flat fronted four-story home with at least 7ft tall gates and a sweeping gravel drive.

'It is a bit big for me now.'

I wanted Madeline (Mado) Bichue to put me in her will.

We entered into a large hallway, the ceiling taking in the upper floor. I love looking at how the other half lived. Mado sat down as we were invited to look at all her photographs on the walls and listen to her life story.

And what a story.

She was married to a Captain who travelled the world on boats, while she was left at home. Then, when she was still young, he died. It was then she decided to buy a boat and make some money in the process.

She bought a sand barge and gathered crew from all over the world. Ethiopians, Moroccans, Russian sailors all made her business work, and they became life-long friends in the process.

'There was quite a bit of drinking of vodka,' she said. Mado took part in the Second World War, ferrying French fighters to safe harbours, smuggling food and wine, plus ammunition.

'There were a few 'close calls', she indicated with a wave of her hand.

We looked at the pictures of the crew on her boat. They were all fit young men.

'Did you fall in love, just a little,' I pointed to one particular sailor.

'Oui,' Mado wagged a finger at me. 'Just a little.'

'And this?' Boomie pointed to a boathouse wheel which looked too big for Mado.

'It is from my boat.' It took pride of place in the hallway.

'And now?'

'I have letters from all over the world from my boys on the boat.' We were led into the huge kitchen, furnished and never touched since the early 40s. There was a bundle of letters with stamps from far-flung places. I decided I would write from Australia the next time I was there.

'And your family Mado?'

'My son, he will have this,' she stamped her foot on the floor.

We were treated to a small tour of the house and the walled garden beyond, which had a commanding view of the inlet and boasted a dock.

'The oak tree,' Mado pointed to a huge stump in the middle of the yard. 'It fell. It was more than 300 years.'

I'm fascinated by that sort of fact. To think of the things it has seen. Amazing.

'We must take tea.' Mado grabbed my arm for the unsteady walk back to the house. Time was getting away from us and as much as I would have loved to stayed and enjoyed the afternoon, Myriam would want to go home.

We promised to return. We promised to write. It was sealed with a kiss on each cheek and we left Tréguier a little happier than when we arrived.

Myriam met us at our arranged rendezvous and we talked all the way home about Madam Bichue, her derring-do, her breaking of the glass ceiling in the 1930s by being the only female skipper in Brittany and her wonderful house.

It was better than tiling, by a mile.

Four

Our Titre de séjour, that (visa) document that lets you stay in France has its anniversary every May. It inevitably necessitated us hanging around for the all-important interview where they look at our credentials, decide we are not going to rob a bank or stage a coup and give us another year.

The rigmarole surrounding getting the interview was prone to change, as the vagaries of the French have no discernible rhyme or reason.

The year before, we were required to present ourselves to the Prefecture to get an appointment. This year we were to email our intentions and get a reply. I duly followed procedure and after a few days I received a reply and BINGO we had our choice of appointment slots. It was so easy I wondered if I'd done it correctly. It was far too easy, and being just a little suspicious I emailed back my acceptance with a reiteration that we wanted a renewal of our visa which ended on such and such a date and our names and our grandmother's names and what we were having for tea that night. It helps in dealing with the French leviathan of bureaucracy to be thorough.

I went to the library and printed off two copies of the email. I wasn't taking any chances, knowing as I did, about the way these things work.

We now had the month of April up our sleeve to do whatever we wanted.

We wanted to go to Le Mans.

Boomie and to a lesser extent myself have always been motorcycle fans. We both had bikes in our teens. We both kept up the love of motorcycling as the years went by. I had only been to one other motorbike event and that was on Philip Island in Victoria Australia many years before, so I was up for a trip.

Boomie had been coveting a trip to the 24-hour endurance motorcycle race at Le Mans for, it seemed, as long as he could breathe.

The motorbike race is held over 24-hours, non-stop, with three riders using the one bike. It has the reputation of being quite spectacular and the bonus that you can watch from any point around the track.

We used the internet to find out all the prices, the options and then it is a juggling act as you book tickets and accommodation all at the same time.

'How about we hire a car and drive there?' It sounded like a good idea to me. We could take in some sights along the way and I'm always up for a road trip. We could do it in one day there, and several coming home. It was only 288 kilometres East. An easy drive for two Australians.

Myriam was, once again, aghast at our audacity, and stamina. 288 kilometres was to her an '*énorme*' (huge) distance. Knowing Myriam, she would pack for a waterbag trip, as we say in Australia. We'd probably take a packet of lollies.

With a plan forming, I went about the process of hiring a car. We opted for the cheapest, smallest car on the payroll and sat back to wait for the fun to begin.

It was the day before we were scheduled to leave that I thought I'd collect all our print-outs, our passports, ID and drivers licences. I spread them on the kitchen table and horrors of horrors I saw Boomie's international drivers licence was out of date. We couldn't renew from France. We couldn't hire a car without it.

'Merde.'

I ran next door and asked Myriam if she would ring the company, explain the situation and see what they had to say.

Myriam thrives on a challenge, and the Ashwins give her plenty of practice.

She rang.
'Non.'
She pleaded.
'Non.'
We lost our money. All of it in an advance payment.
'Merde.'

Now, not only were we without transport, we were without a way to get to Le Mans. I had pre-paid our entrance tickets and our hotel. This was turning out to be a very expensive mistake. We agonised over our misfortune and then I had the idea of the bus.

We'd seen the OUI bus company many times taking people all over the place. It was purported to be cheap, which is always a good option in my book.

I went online and held my breath hoping we could get seats. I mean, how many people would be going to Le Mans via Rennes mid-week?

The bus left at 9ish in St Brieuc and by my reckoning we had plenty of time to catch our local 2€ bus at 6:30 arriving in the big smoke, St Brieuc at 7:30, catch another bus out to the long-haul station and wait. Everything was going swimmingly. It was about 5 o'clock wine time that I remarked to Boomie I'd seen Guillaume, Myriam's son home from school all day and he didn't look sick.

The words had just left my lips and I knew we were in trouble. I pulled out the free calendar from the library and looked at the dates.

'It's the school holidays, isn't it?' Boomie slumped in his seat.

'Yep.'

We knew what that pronouncement indicated. We had been caught once before in the school holidays.

There was only one bus a day in the holidays, at 10 am.

'Merde.'

If there was a God he was trying my patience. Now with one sleep to go we didn't have any way to get to the bus stop.

'Non.'

Myriam was sorry, but she was working early and as much as she would like to help, it was 'impossible' (said with much hand waving and a French accent).

We did know a couple of English people in the village. We had been invited over to their house once or twice. It was a long shot.

Chris said it was not a problem. Don't you just love friends? We promised souvenirs. We promised wine. We said we would meet them at 8 am with our suitcases.

The long haul bus stop is on a round-about and there is a small pull-in bay with no shelter. We said our good-byes and waited with a straggle of other passengers in the cold, all looking down the slip road for the bus to appear. It would be a 4 hour trip to Le Mans, in warmth and comfort. All we needed to do was sit still and watch the world go by.

Now I don't know about you, but it is the Ashwin luck that whenever we have tickets, to whatever event or trip someone is sitting in our seat. We weren't disappointed when we were allocated our seats by the driver, someone was sitting in mine.

It's a sort of French thing, we have found. You have a seat, but you prefer this one, and until someone comes along with a validated claim you just sit and enjoy yourself. I showed the young man my ticket and explained with a lot of pointing, that he was in the wrong seat. I'm a stickler like that. Naturally, all the other passengers were watching my antics. Boomie was busying himself with our backpacks. I looked at the man and he was disinclined to move. I indicated we needed to sit together when the passenger from behind said something and the interloper just bolted up and moved. What the other man said I don't know, but it worked and we were allowed to plonk ourselves down for the trip.

I leaned over to my saviour. 'Merci,'

'Madam, enchanté'

'I think he likes you.' Boomie nudged me. It is nice to be appealing.

The bus stopped at regular intervals to pick up people and for the necessary toilet, coffee, and lunch. I wondered how the French would manage a half-hour stop for the all-important lunch. They took up the *pique-nique* mantra, finding tables outside and a few dare-devils stood about eating sandwiches and smoking. We had packed a cut lunch and so only had to queue for the toilets. Some things are the same the world over. The ladies had about a dozen women waiting, the men were in and out in a flash.

I didn't know where the bus was going to stop in Le Mans, so was pleasantly surprised that it was a three-minute walk from our booked hotel. We spotted motorbikes on the road into the city and as we approached our heads were swivelling around as if we were watching a tennis match. It upped our excitement level a notch, knowing the city was getting in the mood for motorbike racing. Flags lined the city streets, young lads were doing cheap tricks on their bikes, and the police presence was minimal. This was going to be fun.

Our hotel was cheap and clean, but our room was about as big as a man gets for assault and robbery. It had a double bed and we needed to crab walk around it to get to the front door. The shower/toilet were together in a small stall.

'Just like living on a boat,' I said.

'It's cheap.' That was the bonus in the deal. We also found that it was a one-minute walk to the tram station, the last on the line and at the other end was the race circuit. In the scheme of things, we felt our luck was changing. Everything was so convenient.

There was still plenty of day left and a trip into town was only a tram ride away. We purchased 10 tickets for our 5 day stay and headed into the centre for a reconnoitre. The first thing we found was an Irish Pub and a pint of Guinness, which is our all-time favourite alcoholic beverage. They call Guinness 'steak and eggs in a glass', and it is every bit as tasty as the meal. We stayed for two and then thought about going back to our mouse hole - it had been a long day.

You see all sorts of sights on public transport. I watched with interest as a man rummaged through his bag and produced scissors.

'Here we go again,' I said to Boomie as we had seen a man with scissors on the TGV train once before. This man wasn't cutting up magazines, but a sausage. The French tut-tutted and turned away. We couldn't take our eyes off the fellow. He smiled and I said,

'Bon appetit.'

French people are very reserved in public and it's rare to see a commotion on a bus, a train or people doing the wrong thing. Everyone is polite, well-behaved and if a child is crying, the mother curls up in embarrassment, or if a phone rings there are a lot of hard stares. To stare at the scissor/sausage man would be bad manners. To say something, well it just wouldn't happen. Of course, I might be oversimplifying things, but this is just what we have observed, for the Ashwins have no compunction in looking, although we never stare. Well, not much anyway.

We couldn't miss our stop as we were the end of the line and so it gives you time to relax and watch the city light up for the evening. A quick trip to the supermarket for the essentials and we could have an evening in, watching television with our feet up. The shop didn't have anything in the way of quick and easy. We usually go for the microwave 4-minute meals. Rip pour, stir heat and eat.

'I guess we will just need to eat filth.'

Crisps, chocolate and apples and a hot chocolate from the machine downstairs. Oh well, I suppose we don't need vegetables every day. We were in Le Mans for motorbike racing. Vegetables were the last thing on our minds.

The Ashwins are early starters. We have always been up at the crack of dawn, so, it was no surprise we were first on the tram platform.

This was practice day. This was our day for getting into the swing of the race, where to go, what to see.

The first thing to see was the museum. It is full of cars that did the race, and some bikes as well. It is situated right at the race track and has a shop attached for all the 'stuff' that people buy. Key rings, t-shirts, postcards and the like.

It is sort of a one-way wander so you don't miss anything in a back room or off to the side, and all the labels were in French and English, that was a bonus.

Some French museums we've found and some people are incredulous,

'But, you don't speak French?'

'Non.'

They just can't fathom it.

There were a lot of older model cars, cars owned by famous people, famous movie stars, famous racing drivers. It was interesting without being boring. I get no thrill from reading about 6 cylinder, overhead cam, double something or others and triple whatchamacallits. I do like stories about people. The designers, the obstacles they overcame, the partnerships made and lost and the museum had plenty of that sort of thing to make it interesting for me. The cars were accessible too, so you could peer in the windows, look at the plush seats and imagine climbing into a small racing cockpit.

I got the feeling for the place as I read about the wins, the deaths, the death-defying feats. Le Mans has a history of danger. I suppose all motor-sports do, that's what makes it that little bit more exciting.

After a good couple of hours indoors we walked the track, then looked at all the stalls, snaffled a programme and watched some practice sessions. It was a small introduction to the main event the next day. The whole thing was well organised, but a lot of the stalls were shut for the practice session and would only open for the main event on Saturday. We called it 'scoping the joint'. It is always a good idea so we don't miss anything. I can't help it. I like to get my monies worth.

Travel broadens the mind and the hips, so they say. We left the circuit then took the tram back into the city centre for a meal out. What we found was an 'all you can eat' Chinese restaurant.

'That sounds like a challenge,' Boomie said.

'Game on.'

We were early and actually the first in the restaurant so didn't have anyone to watch to see how this thing worked. You never quite know in a foreign place what the protocol is regarding eating, serving etc. Luckily the Chinese lady saw we were English and mimed what we were to do.

They had a help yourself affair and also a pick the raw ingredients and the chef will cook it up. We went for the already cooked. We don't usually go out for dinner; it had been years since we'd had Chinese and this was so good. On the boat we just couldn't be bothered with the dinghy ride home in the dark. We actually lost our yacht once, in the dark, while staying too long at a pub on the foreshore. Eating until you can't eat any more is a rare feeling, but before we'd gone for the dessert trolley we decided we'd come back again.

'I like Le Mans.'

'Me too,' I said as if it only took a good restaurant and a tram ride to persuade me.

Saturday is race day. We trooped to the gate with all the other fans in the early morning and you could hear the engines roar before you could see them.

'I think this is gonna be awesome,' I nudged Boomie as we showed our 3-day passes and our newly acquired wrist bands to the gatekeepers plus had our bag searched. People were arriving on bikes loaded with camping gear. People were walking, scootering, bicycling and we all had that look of expectation in our eyes.

At 2:30 the race would begin, but there was plenty of things to do until the hooter. We walked right around the Bugatti track, scoping out all the various angles, where the public toilets might be and the coffee.

Le Mans is famous for the car race and Steve McQueen made a movie about it. That track is 8.5 miles long, but the

Bugatti motorbike track is only a small portion of the whole. It is 4.185km, an easy walk.

There is plenty of public access to the trailers where the teams have their mechanics, their eateries and their women. We gawked at the money being spent on fully kitted workshops and the catering. It was big money.

If you are lucky you might bump into a rider buzzing about on a moped, or stopping to sign someone's helmet or jacket. There is an easy commraderie about the place and not for one second did I feel out of place, threatened or in danger of having my purse stolen. There is a, *let's all just have fun* atmosphere that has everybody smiling.

Around 1 pm thousands of people start to filter to the grandstand area and the straight for the start. I was amazed that all those people could fit and that we could all see. We were in the cheaper seats, aka standing with the hoi-polloi and could see the whole thing on the tiered steps.

Boomie was wearing his Australian jacket on with the word Australia embroidered on the arm. He purchased it in an emergency for $10 when we were in Sydney getting our visa's from the French Consulate. We'd come from tropical North Queensland and didn't expect to be cold and Boomie bought the first thing he could find from a market stall.

Advertising our nationality isn't our preferred option, but it did garner some remarks from the French.

'Crocodile Dundee' someone said, and another fellow told us that Broc Parks on the Yamaha Austrian Racing team was Australian.

The other thing people remark upon is Boomie's large moustache. The French don't do much facial hair, and they love his moustache. He gets lots of handshakes, slaps on the back, and smiles from complete strangers.

'Bonne moustache.' I think they see it as a mark of a real man. I love it too.

We settled in to watch the race. It is a tradition that the riders run to their bikes for the start. So, the machines are on the inside of the track facing out for a clockwise direction and the riders in their leathers and helmets are on the other side of the track.

All the riders are young fit, skinny lads. Motorbike racing isn't an old man's game.

The master of ceremonies announced all the national anthems and as each was played the crowd remained observant and dignified, and then ... they played the French national anthem, La Marseillaise.

It was a stirring sight and sound. The whole crowd, and I mean everyone, stood and pulled their hats off and sang. You could feel it meant something to them. I was in awe at the patriotism, the verve, the singing from the heart.

Flares signalled the end and a cheer went up as the commentator ramped up the excitement talking faster and faster.

A hooter signals the race to start.

We were lucky enough to be near the favourite, Kawasaki number 11 and I could watch as he raced to his bike, hopped on and was off, only to be passed by Suzuki number 3, before 50mt.

It doesn't take long for the bikes to do a lap and we had 24 hours of non-stop racing ahead of us, so we walked off to visit all the shops on the way to find something to eat.

These events are peppered with merchandising stalls. You can buy anything from a blow-up hand to a brand-new motorbike. We skipped on the latter and became merchandising sluts.

We were seduced into buying a hat, t-shirt, key-ring and other stuff we didn't really need.

We have never, and I mean never bought into the logo frenzy. I had a Golden Breed shirt once in my teenage years, but we were never seen sporting a brand.

I don't know what came over us.

Second childhood by the look of it.

There were heaps of displays of bikes from all the major brands and I saw the very bike Valentino Rossi aka 'the Doctor' rode in a GP. I sat on a new BMW and generally had a ball. Who is to know we didn't have two brass razoos to rub together. We could've been multi-millionaires on the look-out for a new bike. I had a giggle when the salesman went into his big spiel about the bike to Boomie knowing Boomie couldn't hear him over the noise of the music and if he did manage to hear, he couldn't understand. He has a way of looking interested, while his mind is elsewhere. I've seen that look before. I knew he needed saving and stepped in to pull him away.

'I thought you'd never come,' he said to me later.

'I saved you, you owe me.' Fair trade I reckon.

After shopping, we went for eats and decided on a hot roast roll and chips. I tried to assuage my guilt at eating more filth, by telling myself it was a once in a lifetime thing.

Beer was next. Well, you need something to wash it down and we weren't driving anywhere.

The sun came out for the afternoon and people began to strip off in the 'heat', around 22 degrees.

It's only when everyone is equal in t-shirts that you realise there are all shapes and sizes all over the world. The French like their fries as much as the Swiss, the Albanians and the Inuit. The Ashwins felt right at home.

The thing that surprised me at the event was the cleanliness. All the toilets were serviced constantly and some had hot showers. All the 'free' drinking fountains worked and were clean. The grounds had rubbish bins every 100 metres or so, so there was no excuse. It was a pleasant surprise. And the programme booklet was free too. It really was a top-notch affair.

The Bugatti track is on a sloping hill, so you get some vantage points where you can see the riders coming. There are free grandstands with huge screens facing the crowds that way you never miss a spill, or a pass. But oh, the noise. It is a real visceral sport. Some spectators wear hearing protection, we like the noise.

I like spectator watching too.

We were in the grandstand when a very drunk man was trying to climb the stairs. He wasn't going to make it and was swearing loudly. People were laughing at him behind their hands and he made it to a seat next to me and plopped down. Why me? Why is it always me?

'Bonjoooooooooouuuuurr,' he said.

'Bonjour,' I said and pretended to read my programme.

I just knew he wanted to strike up a conversation with me as it felt like everyone was watching. I produced my 'get out of gaol' phrase,

'mon français n'est pas très bon.' #my french is not very good.

I added, 'l'Australienne,' for good measure.

'Ah,' he leaned in close. Garlic, and whisky by the smell of it. Boomie tried to stifle a giggle.

'Crocodile. Snap, snap.'

'Oui.' I made a show of looking at my watch. I wanted to stay in our seats, but I didn't want to be next to this man.

'C'mon.' Boomie came to my rescue and we moved.

Sometimes I feel like a magnet for all the oddball people in the world. It happens all the time. I think I have a face that looks inviting, or it may be that I stare too much.

We found a spot at another grandstand to watch, but it was what was going on behind us that took our fancy.

Le Mans caters for the camper too. The camping ground was awash with colourful tents, thousands of bikers and campervans. They were having a good time by the look of the burn-outs, the roar of engines, the shouts, and the smoking bonfires.

'We gotta come again.'

'We gotta camp.'

It sounded like a plan.

It didn't take much persuasion to go back to the Chinese for a second time. We ate our way around the bain-marie and then back to the mouse hole and a welcome rest.

Our last day at the track was terrific. There was an electricity in the air as the race came down to the last hour. Everyone was track-side at the grandstand for the finish. We mingled as the minutes were ticked off. The compere was superb as he spoke French then English, although trying to understand what he said via a loud speaker was the hard part. It sounded like a train station announcement through a roll of newspaper.

The crowd was thick at the finish line and as the Yamaha rider, Mike Di Meglio came over the line there was a huge cheer. Then, the crowd clapped every rider, right to the last, which I thought very sporting of them. It felt like they were applauding to thank the riders and teams for a great event. And it was.

The surprise happened when they opened the gates to the track. We surged forward with the masses and I had to stop myself from running to the podium near the timing billboard.

'Steady Hettie.' Boomie was walking fast.

I could say I nearly got sprayed with champagne, I was that close. We got some of what was said, and gleaned the rest with the free WIFI, called Wiffy in French.

The winning margin was the closest in the history of the race. The race saw the greatest number of laps in history with 860 completed. And our Australian, Broc Parks was third.

Don't ya just love the internet.

I'd given us a day or two to see the city apart from the racing, so we caught the tram into town the next day and acted like tourists.

Le Mans has quite a long history. There is a fabulous old part where you can walk the cobbled streets and marvel at the lives these old half-timber houses must have seen. There are Roman walls, bits of Roman bricks that are still visible and mark the foundations of houses which are lived in today. Some of the old part is built on a hill and the more expensive old houses have walled gardens with views of the Sarthe River and little flowered balconies jutting into the lanes.

There are so many nooks and stairs from the Roman era, it is easy for the imagination to run wild. We wandered to the cathedral, then took our lunch in the square past a viaduct of enormous proportions. I could see why the Roman's wanted the city, it's an elegant setting next to the river, although we read in ancient times the river was the home of the tanneries. Not the most appealing bit of real estate with the smell and the mess.

Le Mans fired our taste for history, our love of churches and adventure.

We came home on the bus, reliving the race, the track, our merchandise purchased and the Chinese restaurant. We also came home with a promise to go again.

And I don't need to tell you someone was sitting in my seat on the bus.

Five

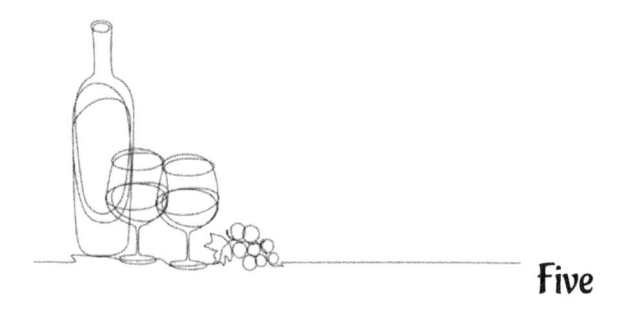

All good things must end, so they say. Our yacht was still moored in Australia and we needed to go back to look after our home of the last 13 years. You can't just lock the door and put the key under the mat. A boat has needs. Its needs were becoming a millstone around our neck and the bank balance.

Originally, we had the idea that we'd sail to France. It was a fine idea, full of adventure, fun, excitement and others had done it, why not the Ashwins. It all started to go downhill when we found that the French would like a good portion of our income for importing a boat. To visit is fine, and you get a full three months to sail the waters. Bringing a boat home for good was a different story altogether.

Plus, they wanted another portion added as duty on all the victualling, the fuel and the whole expense of getting to France. It would cost a motza, which we didn't have, weren't likely to get, and it rankled that what we did have, someone wanted their share. I don't mind paying something, but not everything.

We needed to seriously think about our options vis-à-vis our Australian home, aka Dikera, a 45-foot aluminium sloop double-ender.

With our visas taken care of, without any hoo-ha this time, we booked tickets back to tropical far North Queensland, Australia. Another long-haul flight to add to our growing tally.

Boats are like babies. They need constant care and attention. They also need their bottoms cleaned regularly. We had to do our hull and booked into the hard stand to throw more money at our millstone.

It was 'on the hard' as they say, that a man came up and admired our boat.

'Been lookin' for something like this. Any chance ya sellin'?'

'No, not really.'

The thought hadn't occurred to us. This was our home. This was all we had in Australia. In Far North Queensland, the waters are warm and any antifoul has its work cut out just keeping up with the growth. We decided the thing to do would be to go south.

I rang around and found we could moor *Dikera* at a private dock in Mooloolaba, Queensland, for a reasonable amount of rent. It would mean a sail down south, but the bonus would be we'd be out of the cyclone zone, and there was ample security. It sounded like an opportunity to park up and know the boat would still be there when we got back, if we ever got back. The way I was feeling about our little house in France, I was in love.

Sailing south, in fact, sailing anywhere requires planning. We were told we couldn't have the dock for 6 months. It wasn't ideal, but it suited us as we'd have time to go back to France.

The woman at the Prefecture in St Brieuc had said, while she wagged her finger at us, if we wanted to have a visa, we really should stay in France for 6 months at the minimum. I was all for following the regulations.

So, we cleaned *Dikera's* bottom, put on a fresh coat of paint and moored her in the mangroves in Queensland, paying a man to keep an eye on her for 6 months. Our man, Kenny, lived a frugal lifestyle that included drink. We paid in advance and before we'd even left the dock he was at the bar. Kenny's boat had sunk more than once and been resurrected, it should have been called Lazarus.

There is a crowd that live on boats that are predominantly single men with not a brass razoo between them. There was cookie, boots, bugs, and others. They all sport the 'alternative' lifestyle that includes scrounging, scabbing, and making do with whatever comes their way. One enterprising bloke got a hold of some free paint. Overnight his boat was painted bright orange.

'I like orange,' he said to me as I stared at the slap dash job. Boots liked nothing better than to parade around naked on his deck, just as a dinghy went past, or a tourist boat. I didn't hold out much hope for the care and wellbeing of Dikera, but sometimes you need to put your trust in people and hope that small shred of decency will rise to the top. After all, his one job was to check the bilge once a week. That's all.

'Sure ya not sellin'.' Our little man had a tinny and was going past Dikera.
'No.'
'I'll just give ya me phone number.'

I took the number and filed it away.

We could be back in France in a few weeks.
I scratched another long-haul flight tally marker on my suitcase.

And a Chinese woman was sitting in my seat.

Six

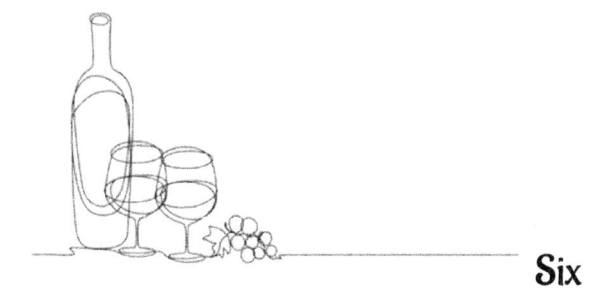

Coming home, for that's what it felt like, we'd decided to stock up on things we needed. Our suitcases weighed a ton as we brought little things we couldn't do without. New sheets, new underwear, those little vegetable peelers that come in a pack of 5 that the French just don't sell. We looked like we were setting up a market stall when our cases went through X-ray.

The flight was via Dubai with a 22-hour stop-over and it was an easy decision to have a hotel room, see Dubai and get ready for the next 14-hour leg of the trip. I can confidently say I don't like long haul travel. I did enjoy Dubai, albeit a short stay.

The icing on the cake was my trip to the Mont Blanc shop in a huge shopping centre. I'd promised myself when I reached a milestone in books, I'd treat myself to a MB fountain pen. They are the luxury, top of the line in pens. The MB shop, that hallowed ground where customers are well-heeled was suitably quiet. Once I indicated my choice I was given the Royal treatment with coffee, my own assistant, ink to try my pen, nice paper, and made to feel like I had a million dollars up my sleeve.

I came out walking on air, with a pen in a box in a cloth bag in a box, in another bag and in a shopping bag, all with the Mont Blanc logo. I could get used to high-end shopping if this is how you are treated. Boomie and I *felt* like millionaires and I guess that's the whole point of the experience.

We left Dubai with great memories of food, the waterfront boats, the souk and just missing out on a carpet with superior quality that the man assured us we'd have for many years. He was good, but not as persuasive as the Mont Blanc assistant.

Lugging our cases, sans carpet! up the hill from the bus stop was no mean feat, but the thought of putting our feet up, buying a bottle of red and turning on our little heater sustained us.

With a foot in each hemisphere, it should have been hard to put down roots. But we felt right at home in our little village. I guess that is because things felt familiar - because things never change.

Oh, there is the odd building renovated, the shop-front awning finally collapsed, but the spirit, the essence of the place remained the same. It was a small village in rural Brittany and things carried on the way they always had - with slow, easy, familiar regularity.

That sort of thing can get under your skin. It's addictive and attractive.

Life is governed by the seasons, the Saints days and the growing cycle of the little vegetable patch everyone had in their yard.

We came back to winter vegetables and a cold snap.

There is a local market that our neighbours go to every week and one morning when Boomie was out the back Marie-France said something to him and being a little deaf, Boomie nodded and said,

'Oui.' It usually gets him out of trouble.

We busied ourselves walking to the shops only to find when we came back Marie-France and her husband waiting in their car, outside our house. It turned out Marie-France had asked Boomie would we like a lift to the market. We rushed inside, deposited our shopping and apologizing profusely sat in the back of the car. How long they had been waiting we didn't know. They were gracious and friendly and didn't mention the incident. I bought them a pineapple and some grapes, those things that are a bit more expensive and not readily available in winter. You can't go wrong with fruit.

Things were looking permanent now we had our boat sorted, albeit for 6 months. Another long-haul flight, but maybe the last for some time.

With a feeling of 'we are going to make France our home,' we decided to buy some transport. You don't realise how convenient a car is until you don't have one. If I lived in Paris I probably wouldn't bother, cause there is nowhere to park anyway, but in a rural setting, a car is bordering on necessity. We wanted to go places, see things and camp.

I had found there is an organisation that you can join whereby you can camp in any vineyard as long as you buy a bottle of wine. I mean, if that isn't a match made in heaven I don't know what is? But camping requires 'stuff'.

I went on the internet in search of 'stuff' A tent was top of the list.

I'd done a bit of homework in this regard. At Le Mans we hung about the camping site and looked through the fence at the different tents. I took photos of the brands, the designs and noted the popularity of each type. So, when it came to searching the internet, I knew what I had in mind.

We also knew what we had in mind regarding transport. We wanted a van, newish if possible. All that 'stuff' had to go somewhere.

The Ashwins are haul-it, carry-it, dump-it people. A sedan just wouldn't cut the mustard.

We took ourselves to St Brieuc to look and talked ourselves stupid. We said how we would be sensible. How we will not buy the first thing that comes along. We weren't born yesterday you know. Words like canny, street smart, wise, eye for a dud, a lemon. We can spot a shonky deal a mile off with our eyes closed. You can't pull the wool over our eyes. We were so ready to hunt for a car or van we were bordering on black belt experts.

We hopped off the bus and walked down to the Citroën dealer to scope out the forecourt, but within the time it takes to brush the crumbs off your sleeve we were sitting in the office filling out forms.

'I'm pretty sure we did the right thing.' Boomie said on the way home.

We congratulated ourselves on decisive action. We told ourselves, what's the point of traipsing about the place when we were smart enough to take an opportunity when it presented itself. You know you can convince yourself of anything if you try hard enough. We'd had years of practice at this sort of thing. We thought ourselves clever beyond measure.

We had bought a Berlingo van. A second-hand almost new white van with 153km on the clock. It had a GPS and all the whizz-bang stuff. It had enough room for camping things, renovation things and it was diesel. We could, if the mood took us, go to Paris and back on one tank of fuel. I liked the sound of that.

We took the bus back the next day with all our documents to clinch the deal. The documents were a doddle because I had learned early on to keep everything about our sojourn in France.

The one thing everyone needs is proof of residence. What you do is email the electricity company and they shoot an email back saying who you are and that's where you live. The *attestation* is essential in any transaction. It also needs to be less than three months old. I felt like an old hand as I applied and received my document. I forwarded it to the dealer so he could print it off. We walked into Michael's office and sat down to talk turkey. Michael was the only one in the place with a good smattering of English, which was a Godsend. We explained that we would pay for the car straight away.

'Now?'

'Oui.' I couldn't see the problem. He couldn't quite believe it.

'Cheque.' I put my cheque book on the table. Michael thought it was a joke. I don't know how they do things in France, but we just wanted to pay.

'Who are these strange people,' I heard him say to the office lady, who was the financial controller. We were creating quite a stir in the place. The manager came in to look at us.

All we wanted to do was pay.

'No. Sorry.'

I asked for Google translate and typed, *we would like to pay the full amount.*

The manager shook our hands and smiled. I figured most people lease their cars, or put them on terms. Here were people who wanted to just 'buy the damn thing'.

'Extraordinaire.'

The first hitch came with a cheque. They didn't take them. All over France people do business with a cheque book. From groceries, petrol, hardware, everywhere they take cheque. Citroën dealer?

'Non.'

Now what?

'How about a bank transfer,' I said in my best French.

'Now ya talkin'.' Michael nodded and smiled. The manager nodded and smiled. Madame financial controller nodded and smiled.

There was just one tincy wincy little hitch in the deal.

I'd never done a transfer from our French account. I'd done heaps of online transactions in Australia, and knew the drill. I knew all the French words for transfer, deposit, transaction, and all that sort of thing, so

'How hard can it be,' I said to Boomie.

Michael logged on and passed the computer over to me and I logged on. Lucky I knew the password off by heart.

I tried with all the numbers Michael had given me and the thing wouldn't go through. I could see our healthy bank balance, but the transaction was a dud.

I handed the computer screen over to Michael and he verified all my numbers. Zip. Nada. Nothing would work.

It was then Michael saw a little caveat that said something that put a dampener on the whole thing.

Apparently, we needed to ring the bank and let them know we were adding a new payee to our account.

I groaned inwardly, and outwardly.

'What?' Boomie looked on.

'Well …' I would need to ring Paris because that was the branch we did business with and explain that I wanted to buy a car and that we wanted to add a new payee.

I Google translated the whole thing knowing full well the Paris office were finicky at the best of times. Several times I had rung only to be disconnected when no-one could understand me. The word snooty came to mind.

Michael was pulling out all the stops now to get the sale happening. I handed my phone over to him after I had dialled and he went into overdrive explaining the situation. I heard Australienne quite a few times.

They hung up on him.

I didn't want to say, 'I told you so,' but it was on my lips. I really should learn that phrase in French, if only to say it under my breath.

We all broke for coffee and the financial controller came in with cake.

Michael tried again after I had dialled and we managed to get someone a bit more agreeable this time. It was arranged that I had to recite my acceptance for Michael to act on my behalf. I passed the test.

How hard is it to spend your own money? We did the transfer and there was a great whooping and congratulations, felicitations to all.

'I think we just bought ourselves a Berlingo van.'

We expected to collect the car in a day or two, nothing much to do except a registration transfer.

'This will be our last bus ride for some time to come,' I said on the way home.

We celebrated with a bottle of champagne and wondered if the van would fit under the house.

Our house is built into a hill so there is the same room under as above. The only problem we could see would be the doorway and the middle overhead concrete beam.

The doors were 10cm more than we needed after we Googled the dimensions including the sticking-out mirrors. Our little van didn't have the folding mirror option. It would be tight. As for the overhead beam, we had enough to get the van in with just enough spare to close the doors.

So we swept downstairs, made space and looked forward to having transport to tootle about the place.

Michael said he would ring and email. I like email because it has that translation facility so there can be no miscommunication.

We went to bed full of the things we would do, the places we could go, the 'stuff' we could carry.

I carried the phone with me constantly for three days waiting for our little van.

Nothing.

Then it was the weekend. Nothing happens on the weekend.

A week and I emailed Michael. His message pinged that he was out for a few days.

Myriam, our neighbour said in her inimitable way,

'It is usual.'

We bought a bottle of wine and waited.

After 10 days we were champing at the bit. We started to feel ridiculous waiting for something we had paid for and were not getting.

I sent an email trying to get some definite date when the phone rang just one minute after I sent it. Co-incidence? Hmmm.

'You can come.'

We caught the bus the next day and walked to the dealers and saw our little van was waiting for us on the forecourt. It had been washed, polished and had half a tank of fuel. The registration completed too. In France, you only pay once for registration. That's it. Once, for life. Everything was as it should be and really, I don't know why we were so impatient. These things take time. There are procedures for everything. Ten days isn't too much really.

As I said, you can convince yourself of anything if you try.

Michael showed us the features, told us to drive on the right and waved us off.

We knew the way home by bus which takes in all the villages, the twists and turns, but going direct was new and exciting. We saw a chateau we didn't know was there. We realised there is a shortcut from one village to the next and we also went shopping. Being able to carry more than a back-pack worth of goods was a real benefit. We bought fuel for the little heater too . . . and soon discovered the wood flooring in the back of the van was like an ice skating rink. Things slid all over the joint as we negotiated the roundabouts to home. I could imagine the headlines

Australian couple blow up on roundabout with three containers of heater fuel.

Living on a boat I knew a thing or two about non-slip. What we needed was a non-slip mat in the back. That stuff is like superglue.

We drove into our driveway like conquering heroes and waved to the neighbours.

We had a van.

And our tent arrived in the post.
Now ya talkin'.

Seven

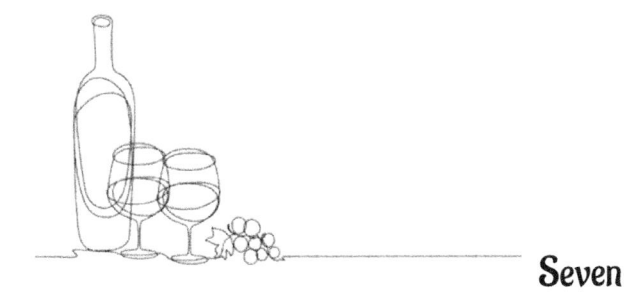

Our next purchase was a motorbike. Boomie had been studying the form guide for some time and knew what would suit us to tour around and what we could afford. The money had been in the bank for ages, just waiting for the day. This was the day.

We'd already scoped out the various motorcycle shops in St Brieuc. Window shopping is the easy part. In France, it is called *lèche-vitrine* which translates into licking the glass. I like that. We did quite a bit of licking as we priced various models of bike, deals on offer and availability. We knew if we travelled to another city we could get something a little cheaper, and that is always a good bargaining chip. Boomie isn't a bargainer, but I can haggle with the best of them. It's like a sport with me.

We had our eye on an FJR 1300 2016 in the local Yamaha dealer's showroom. It was being offered a little cheaper as it was an older model. So we went in to talk up a deal with Michele.

He didn't speak *any* English at all. He could say AC/DC. That wasn't going to help. I took out my phone and Google translate became our best friend.

Now, I have found there are some people who just have the knack of extrapolation when it comes to getting the gist of what you want to say. They sort of get it right off the bat.

'Oh, yes, I see what you mean,' type of thing. Michele wasn't one of these people.

It was hard work.

We said we would like to buy this bike, and pointed to the silver FJR.

'Oui.' A sort of, don't we all nod came over him.

'We pay now by cheque,' I said. I lived in hope that the Yamaha dealer wasn't like the Citröen dealer. You just need to be an optimist sometimes.

Right off the mark I could see Michele, despite his female sounding name thought Boomie should be conducting the negotiations. It was that sort of frightened look Michele had as he glanced from Boomie to me and back again.

Boomie saw it too and explained that he was a little hard of hearing. It always pays to have an excuse up your sleeve for just such an occasions.

I pulled out the cheque book and that threw Michele into a spin.

Here we go again. I could see in France, people just don't walk off the street and pay for things. He ushered us into his office and we sat down. We waited for the coffee from the secretary/financial controller.

'Cake?'

'Non merci.'

'Please,' the financial controller giggled. It was her only English word and she wanted to use it.

'Please.'

'Thank you very much.' We took cake.

Michele apologised for a messy desk. He began to tidy up as we waited and sipped our coffee.

I fleetingly thought it was lucky we didn't have a bus to catch. This could take all day. But it wouldn't take all day, because things close at 12 and being a Wednesday they wouldn't reopen.

We were shown Michele's office chair, which was like a racing car chair. We were offered a sit in it, to get the feel. I pushed Boomie for the tryout.

Then the woman came back with a pamphlet. So that's what we had been waiting for all along.

We admired the pamphlet. We'd seen the pamphlet before on the internet. I'd downloaded it to my computer and we'd pawed over it more than a dozen times.

'Nice,'

We were led out to the bike and all stood about looking at it.

I began to wonder if we were on the same page.

'*nous aimerions payer pour cette moto aujourd'hui.*' We would like to pay for the motorbike today.

That got him moving. We adjourned to the office and he began to fill out forms. We had all our documents in order and Michele was impressed. You need to get up pretty early to outwit the Ashwins. We'd negotiated a house, bought visas and just purchased a van. I'm all over it.

He looked at me with an expectant air. I whipped out my cheque book.

'D'accord?' 'Ok?'

'Oui.'

How easy was that.

I filled in the numbers and then copied the written numbers and signed. Boomie signed. I put our address on the back. I took a photo of the cheque. Knowing the vagaries of the bank I never leave anything to chance; anything could happen and sometimes did.

We were given a tour of the workshop and shook hands with all of the mechanics. I speculated if this was usual protocol.

'Australien.'

Ah, now it all made sense. We were unusual. How many Australians come into the local Yamaha dealer in St Brieuc and buy a bike. Not many I'll wager. A mechanic hopped around like a kangaroo and said 'kangaroo,' which brought the workshop to standstill as the lads fell about laughing.

I wondered if we met the French in Australia and mimed a guillotine would it have the same effect? We were offered another coffee which we declined and Michele told us the dealer service would be done and then he would contact us by phone.

Naturally, I tried to put a date to the ringing bit of the deal, but Boomie didn't want to get into pulling me off Michele as he pulled a date out of his arse. We left in hope we would have our new bike within the week. It was new, so we thought it didn't need all that extra registration transfer etc stuff. A number plate, a signature and 'sweet as a nut'.

While we waited it snowed and Boomie confessed to being a snow virgin. Your first time is always a little special. It doesn't snow very often in Brittany; rain, yes, snow? Well the last time it really snowed was in 2010 so Myriam said, as she watched the Australians cavorting in the snow. Our house looked quite lovely in white. It made you feel like you were in Northern Europe in a Hallmark card.

We have snow in Australia. Lots of snow in some places in the Mountains. They are called 'the Snowy mountains, the Snowy river,' but to see snow in Europe brings with it a feeling that those Christmas cards weren't lying. We sat inside and watched the flakes build up on the fence posts and made sure we had enough heating fuel in case the roads closed. It is the same the world over. A little adverse weather and the place becomes paralysed. It is like it has never happened before and caught the municipality on the hop. Overflowing drains? Well that's never happened before. Snow sweepers? We let them go to the next district about six years ago. Gritting? I know it's around here somewhere.

We took a walk and our village looked pristine in the snow. Snow makes everything look clean and sharp. We heard our first crash about an hour after the fall.

I should point out that we live opposite a crash repair place, so we see all the mishaps. The *Carrosserie* would be rubbing their hands together as the weather closed in. We were not going to chance driving in the snow with a new car. I watched for the grit spreader and a chance to get out and about to see the countryside.

It came on day three with a bit of a thaw. We bundled up a flask of coffee and went out to see what we could see.

I cautioned Boomie on driving on ice and snow. It can all be going swimmingly, then you lose traction and you are sliding with no hope of stopping. We inched our van along until we were on the main road and once there, the trucks had done a good job of clearing the slush. We approached the roundabout with extra care as great chunks of frozen slush were strewn across the intersection.

There was a close call when we lost it completely on the exit and ended on the grass. All a learning curve. And we became overcautious as we practiced our road rage and indignation at the others on the road.

'He's just going too fast.'

'Look at that. Pretty damn stupid.'

'What are those people thinking?' Of course, I think the other drivers' were practicing their swear words at us too. 'Fuck' looks the same on everyone's lips.

The countryside was wonderful in white. Snow highlights the majesty of the oaks; the starkness of hedgerows and you see little tracks of animals.

I suggested a walk in the woods the next day.

We have a huge forest within a ten minute walk from our house. There is a well maintained track which can lead you on an 8km walk, or you can do a logging track and end up back at the village with a good cardio workout. The logging track is almost a perpendicular dirt race. We'd seen horse hoof tracks on it many a time and other tracks of fitness fanatics with dogs, but we were just plodders and mouth breathers. There is also a grand farm house cum mill with a water race, a pond and a weir. The cold snap had frozen the mill race and the icicles were clinging to the rocks, in spectacular fashion. The ice on the pond was a good inch thick and the pine trees all carried a dressing of snow on their branches.

We skittled rocks on the pond making weird alien noises and poked sticks at icicles.

There were tenacious blackberries still clinging to the bushes feeding the birds and little creatures and we saw tiny paw prints all over the ground in the brambles. You can see why faerie tales come from Northern Europe. There is just a feeling that gnomes, elves, faeries, trolls, goblins and the little people live in the forest. I might have read too many books in my childhood, but my imagination works overtime when I see things that ignite the spark.

Brittany has a lot of granite and in the forest there are huge outcrops and boulders of grey granite. Some have been split for building, but there are others that form, natural or not, archways and standing stones that resemble Neolithic sites. Again I wonder how those people kept warm. Did they range over the area in search of blackberries and squirrels to eat. Boomie and I often muse on the nature of nature and man's place in it. We are not the first to tread the path, nor will we be the last. There is something to be said for the continuance of life.

And as the weather improved a little, we waited (im)patiently for the phone call about our bike. After a week we drove into town to see what might be the hold up, forgetting it was a Wednesday.

I pressed my nose to the window and our bike was still sitting on the shop floor, looking like it hadn't been touched. It was a worry. There was a '*vendu*' (sold) sticker on it, but other than that it looked just as it did.

'Perhaps they are waiting for the cheque to clear.' Boomie had a handle on it. That might have been the problem.

I checked the bank when we were home and the money had gone.

'Do you think I should ring?'

Obviously, there is a protocol about these things. As maddening as it was to sit still and do nothing, we would just need to wait.

It was when I found a mushroom growing in the window frame that we decided we needed a better solution to our heating problem. Our house was so wet all the time, it was unhealthy.

We went shopping for a wood stove/combustion heater.

I should explain that our house is over 70 years old and was built just after the Second World War when things were scarce. The beams under the house are only just up to the job of keeping the floor in place. Some have bent and split over the years and the sag in the floor is so much that you can't keep an egg from rolling off the table.

A wood heater is made of cast iron. Not an insignificant weight on the already stressed beams. We agonised over where to put such a beast. It would need to be the middle of the house for maximum heating effect, but it would need to be as near to the only supporting wall under the house. And then there was the issue of the roof.

Houses in France thrive on symmetry. Usually they have two chimneys, one at either end of the gable roof. We had had two chimneys once, but one was blocked over and the pot on top taken away. We suspected this was because of the water cistern attached to the side of the house foundations. This 'tank' was the catchment for the roof gutters and always had water in it … which was undermining the foundations albeit slowly. This made the wall lean outwards, and when combined with a chimney might have been too much to bear. It was a theory, but a good one. Hence we only had one chimney, but still had a fireplace in the kitchen and lounge, going nowhere.

The other chimney was for our boiler. We would need a new chimney right through the roof. So, the heater needed to be positioned so it didn't fall through the floor, but in a spot where it had clear access to the roof and didn't hit the roof ridge. Tricky. I just couldn't get my head around how they would bring the thing up the stairs. Those things weigh a ton ... literally.

The next village over to us had a decent hardware store that carried most things for the home. Our junk mail said they were having a sale of wood heaters. It was time to go shopping.
But first we called in an expert in these things.
Sean was English and knew his stuff. He put together a quote for a heater that sucked in air, blew out air, had a flue thingo and almost cleaned itself and whistled Dixie.
It was, with instillation more than we could afford. We asked if he would install something we bought elsewhere. You know that face people make when they step in dog poo.
He'd do it, but he couldn't guarantee anything about the heater.
'Fair enough,'

Like anything you buy that is going to cost a fortune you do your homework. I'd looked at Myriam's heater and the mess it made. She also had to replaster because her heater was too hot and the plaster fell off the wall.
I looked at brochures and the various designs. We wanted something within our budget, that would heat our house, and not have a big footprint.
You can get all the bells and whistles, and then you can get cheap.
We went for cheap, and the delivery was within the radius of the shop so it was reasonable.

We waited with baited breath thinking of how warm we would be. How I would clean it every day religiously. How we should buy a wood basket, a kindling basket and matching poking, prodding tools. We are both a bit like pyromaniacs when we see a fire. Our wood stove came quicker than our motorbike.

The fellow that did the delivery was built like Arnold Schwarzenegger, and he craned it through our kitchen window. We felt like pieces of wet spaghetti in comparison. As Boomie and Arnold heft the stove across the floor, I heard an ominous crack as they pulled and pushed it into position. I could just imagine the whole floor giving way as the house threw up its hands in despair. Arnie thought it all great fun and jumped up and down on the floor to show how tough it was. All I could do was swallow hard and hope he'd calm down before he went down, right into the garage underneath.

We had to wait for Sean to do our chimney, but just the thought of a roaring fire was enough to keep us warm. I had ideas of rocking chairs, knitting by the fire, a griddle and plate warmer.

'Settle down Hettie.' The voice of reason interrupted my dreams.

'It's small, it's cheap and we don't have room for a rocking chair.'

'I can dream can't I?'

The chimney for the heater came in segments and I was playfully putting them together and they just slipped in so easily how was I to know they wouldn't come apart - ever again.

We pulled, we tried to twist, we wet it, dried it, heated it and part A wouldn't release from part B.

'What the hell.' Boomie and I sat back with the chimney sitting at the table like a guest and gave up. I was afraid of what Sean might say. I'd seen that dog poo look before.

And finally, just when the weather turned cold and snowy again we were allowed to get our bike.

'About bloody time.'

We had another coffee in the office and then Michele gave Boomie the keys. I was to drive the van home and this would be the first time I'd driven it. The thing is I have a wonky elbow so driving is a pain, literally and Boomie usually does the bulk of it. I *can* drive, and *do* drive, but not often.

I stalled it right outside the shop on a hill and brought Boomie, who was following, up sharpish. Then I got the hang of things and kept reminding myself to drive on the right. Boomie is quite ambidextrous when it comes to driving. I need to constantly remind myself which way to look. Old habits die hard they say. I didn't want to die driving head on into traffic.

Boomie followed me home some of the way, then I followed him. The van said it was 4 degrees so I knew that it would be a chilly ride.

We arrived home to a message that Sean could do our chimney in three weeks.

In three weeks we had contracted a man called Charles to cement under the house. There would be chaos for a month, but then we'd be warm, the under house would be clean, cemented and not just packed dirt, and we could park the bike and car out of the weather. It was all coming together.

Eight

When you get something new, there is the itch to try it out. We were waiting, and itching for a weather break to go out on the bike. We had long underwear, scarves, and gloves and thought it all quite adequate.

It was on Boomie's birthday the sun came out and in a respectable 10 degrees we went out for lunch at a place I'd seen on the map.

Perros Guirec is on the north coast of Brittany and has magnificent harbour views, old houses built into cliffs with impossible driveways and a marina.

It was freezing on the back of the bike. Boomie had the engine and heated grips. I had frozen legs, hands and feet. We bundled into a pub and ordered a coffee then some hot lunch.

I don't know if we have forgettable faces, but the waiter and cook forgot about us. We saw others get served and we sat waiting. Eventually I had to say something, but it was obvious the landlady behind the bar had been sampling the goods. She laughed and threw out an order to the cook with a few swear words and I sat back down to wait.

Our burgers eventually came, and they were worth waiting for. Big, juicy steak, salad and chips and a complimentary beer for the mix up. I didn't want to leave as it was warm, I was full of food and it was cold outside. My thighs were just getting their feeling back.

It seems that whenever we go places we invariably end up at a marina. I guess spending years on a boat rubs off and we like to look at boats. I don't think we are alone in this pursuit.

Perros Guirec has a fantastic coastal road coming in and going out and it was this road we followed to get back to the main highway. The view of the coastline is stunning, with outcrops of granite rocks, cliffs and an unforgiving sea. There are memorials to those lost at sea all over the coastal towns. Fisherman washed overboard, drowned in nets, sunk in boats. The list of accidents goes on for years and years. Everyone, but a widow watches the sea.

Coming home we took the motorway and revelled in the power of the bike, the ease of handling and that magic feeling when it all comes together. That other magic feeling was my feet and fingers going numb with cold and my legs cramping up.

If we were going to get serious about motorbiking in Europe we needed proper gear. The stuff we had from Australia just wasn't up to the job. They say poor people pay twice. Now we went in search of winter gloves, real thermal, Scott of the Antarctic long underwear and socks that had a thermal rating of a hot spring in Japan.

I bought gloves with a fake fur lining and silk inners. I bought a long sleeve underwear thing with a 5 star rating and hot pink leggings that assured me I would be warm in minus conditions. I trust labels as much as I trust a used car salesman, but as the price was twice what I paid in Australia I thought my new undies should be up to the test.

We tried our gear out on a ride to our friends who we'd met when we stayed at their AirBnB.

Cathy and Daniel lived in an old house next to the Nantes-Brest Canal that was damaged in WWII when the Germans bombed the bridge over the canal. They were doing up their house and we'd stayed in contact.

It was an easy ride to their place and all my new gear performed perfectly. I was warm. I couldn't imagine being cold again. I thought I might just live in my long-johns. How did people get by without them?

Cathy and Daniel invited us to wine and cake, bread and smallgoods, cheese and dips. I'd bought some flowers as is the done thing and we sat down with Google translate for a convivial afternoon.

It often happens that the merest of meetings sets a friendship, often for life. Cathy and Daniel and their son were our sort of people. Although we only had a smattering of each other's language we got on famously, laughing uproariously at jokes, and enjoying their warm, cosy kitchen, albeit tiny. Their house is three stories with an ancient staircase, wide tiled hallway and tall ceilings, but the kitchen was just big enough for the sink, and table. Most of the room was taken by an enormous old stove and flue. The sort of thing you would see in a castle. Cathy cooked wonderful things on this old contraption. The detritus of living crowded every surface with a cat or two thrown in which added that homely feel. They were intrigued as to where we'd been, what we had in mind and then Daniel suggested we go to the Valley of the Saints.

'You don't know?'
'Non.'
'It's very famous.'

People always think their local attraction is world renowned, on everyone's bucket list and worth a plane ticket. We'd never heard of the Valley of the Saints in Carnoët.

We said we'd go and as a parting gift, Daniel gave us an old bottle of wine. He said it was special, smacking his lips and raising an eyebrow. We said we'd come again and enjoy it together.

With directions, our wine in my backpack and a fond farewell we set off to see what all the fuss was about.

The road to the Valley is uphill and around sweeping bends that made Boomie crank over the bike and we enjoyed the ride through dappled shade and well-kept roads.

It all ends with a gravel driveway to a carpark on a slope. Not the best for parking a motorbike. I hopped off and scouted around for a suitable spot to park. It would become a joke as I said I walked half of France looking for a park. We ended up over a septic tank, the only available concrete around.

The 'Valley' is in fact, the side of a hill overlooking the countryside. And on the hill are sculptures from granite. Modern representations of saints. If you have seen Jesus in Rio then that is the same representation. Although Brittany has a certain nuance to the sculpture and they looked blockish, and angular. They are quite striking, reminding me of the Easter Island sentinels. Made of Breton granite they stand 3 to 4 metres high.

On this cold day the place was packed with people walking the wind swept site. But the view …

Standing on the top of the hill we could see for miles across Brittany. It was certainly something to behold. The villages and farmland cut across the landscape divided by roads, hedges, forests and rows of trees. You could see manor houses, and it all looked a little surreal, like a toy train set.

There are around 50 statues in the grass and they represent the monks who came over from Wales, Ireland and Cornwall to bring Christianity to Brittany. In time they hope to have 1,000 statues. I can see why Daniel was insistent we go. It is one of those places, and one of those sights that sticks with you.

After a good ride the urge to go again is almost overwhelming. We decided we'd go out the next day to the beach around St Brieuc as someone told us you could get a great mussel lunch right on the seafront. It sounded like a plan.

The plan took a u-turn as our little van decided to play up. A warning light was blinking telling us to add some 'Ad-Blue' that ingredient that cuts down the emissions from the exhaust. Europe has quite strict emission standards and our little van was Euro 6, but in the brain of the car it said we should add some 'blue'. We assumed the mechanics did all the topping up etc so were a little surprised. The main worry is that the van's brain gives you so many kilometres before it shuts off and the word is that it is a devil to get going again with all sorts of electronic gizmos needed to override the disaster. We watched the kilometres slowly dwindle until playtime would be over, so went back to Michael to see what could be done under warranty.

He didn't believe us until we showed him the blinking light.

Of course, we fix.' That was a load off our minds.

'Under warranty?'

'Naturellement.'

We were given another van in the interim and praised ourselves for buying from a reputable dealer. Self-congratulation comes easy with the Ashwins, it stops you kicking yourself for being stupid … most of the time.

Our next bike trip was in search of mussels for lunch. Brittany is famed for its seafood and mussels are about top of the list. They usually bring the steaming pot to the table and you use one half of the shell to scoop the meat out and mop up the remaining juice with your bread. It is all finger food and absolutely delicious.

St Brieuc is built around two valleys and to get down to the sea level you take various winding roads with wonderful views. We made our way down to the marina and meandered along looking for a good restaurant. It was a Monday, so we thought we'd be in luck.

We still had an hour before the obligatory 12 o'clock break and dropped into a small, old, undistinguished pub for a warm-up coffee. Little did we know that we would stay more than a few hours in this place.

The woman behind the bar was Breton tiny, around 4 foot and stood on a stool to pull a pint. We had two delicious expresso coffees and began to chat when a fellow came in for a Pastis, that aniseed tasting drink that is very popular. He joined in the conversation when he found out we were Australian. Then a few more of the local motley crew came along, old salts by the look of them and it soon became a party. We had wine, and then snacks, and there was a lot of hand waving, joking, talking rubbish and back slapping. When they found out we had lived on a boat, we were their best friends in the whole wide world. I always carry a few pictures of the boat on my phone, if anyone asks. They asked. My phone was passed around and we were like family now. More wine was consumed by me, Boomie was mindful of the French police as there are very heavy penalties for drink driving in France. Marjorie showed us her family photos including her uncles who fought and died in the

World Wars, her mother who was over 100 years old and her relatives from the 1800s.

We ordered toasted sandwiches, and heard her story about the bar. She had taken over when her husband died and was an institution in this small backwater. She had loaned money to so many people she was a Saint, the fisherman said. He crossed himself and Marjorie held up a fish kosh in retaliation. It was all in great fun.

We were reminded of a pub in Townsville in North Queensland that had the same sort of clientele. Fishermen who count the pub as home. We'd had many an entertaining evening at the Vic Park watching the drunken antics of men of the sea. One patron was thrown out for chucking a cheese sandwich. The landlady was their surrogate mother. It seemed Marjorie was the same to these men. Sometimes it is those ad hoc moments that are the best. We didn't get our mussels, but we did make some friends, drink some wine and all in the most authentic little bar, far from the madding crowd. That's how memories are made.

After a fine day like that, we didn't want to do anything, but go places and meet people, but the big crunch was coming. Sean was about to start our heater instillation, Charles was about to start our cement floor and we had contracted an electrician at the same time to rewire our house while he could get to all the wires before the cement went down. And we had a phone call from a tiler who said he could come and do the bathroom floor and shower stall.

March would be all go. I'd budgeted carefully and it was all in the plan. We just crossed our fingers there would not be any of those 'unforeseen circumstances' that crop up when tradesmen are involved. In France they are called Artisans, but the real artistry is in the bill.

Nine

Charles was a one man machine. He worked alone and just plugged away at it for his allotted 8 hour day. We had designed to have the cement floor a little lower than the previous dirt so we could drive the van right under the house.

'Not a problem.' Charles said in his broad Cornish accent.

He would greet us every morning with a,

'Owwwright?' and have his morning tea with us in the kitchen.

Sean, meanwhile was scrabbling all over our roof and cutting holes in the kitchen ceiling.

I'm one of those people that like to keep track of what is being spent, what is being bought for the job and how much it all costs. We were on a budget and I didn't want to be left short.

Boomie is more of, 'just let them get on with it.' I really tried to let them get on with it. I really did, but when Mick the tiler said we needed a new 'u' bend because the one we bought didn't fit, I had to say something.

'If he needs it, he needs it.' Boomie brought me up sharpish. I clamped my mouth shut.

You don't want to get the tradie offside by being a shrew.

When Sean said we needed a grate thingy on the chimney, and I'd never seen a grate thingy on anyone else's chimney I ground my teeth and went out into the garden. I'm the same in the boat yard. I just can't stand by and let other people spend my money.

We got the 'u' bend. We got the grate thingy and Charles worked away at it, rain or shine.

At the days end we shut the door and took a deep breath, then ranged over the house looking at the handiwork and wondering if it would all come together. It was beginning to feel a lot like renovating the boat. I could see shortcuts being made. Boomie said it was usual practice. I could see quite a few 'usual practices.'

'Don't worry about it.' His was the voice of reason and pragmatism.

One morning I was just finishing the dishes when the whole floor shook with a huge dangerous sounding bang.

I rushed downstairs expecting to find a broken beam, the house collapsing, but all was normal.

'Boomie?' I asked.

He pointed to the beam above my head. Charles had misjudged the swing of his digger and the bucket had whacked the beam. It had a big gouge out of the meaty part and as it was undervalued to begin with, this was a worry.

'Don't worry about it,' Charles said.

Easy to say when he can go home at night.

Charles would come up the stairs for morning tea and carefully introduce a problem, laying it on the table with his biscuits.

'You know you need to do some'it about that wall.'

Boomie and I would look at one another. Our one internal wall that was holding up the floor was on laughable footings. Charles had an idea to shore up the wall with shuttering and cement.

Put it on the bill.

'You know you really should do some'it about the driveway boundary.'

'Put it on the bill.'

And so it went.

We had our own ideas too. I could see myriad of small jobs that would save us hours of back breaking work.

While we had Charles with his mini Pelle, a tiny excavator that would fit in limited space, we asked for his advice on grubbing out a pampas grass that was out of control in the back yard.

'Not a problem.'

I like a can-do man.

'What about this bush?' I pointed to a scraggly thing that hung right over the washing line.

'Owwwright.'

'How much.'

'Let's call it an hour's work.'

I like a man who can put a price on something without scratching his balls and sucking in air like it's on special.

Then, Charles asked about our drive. To get our van in we needed to be extra careful of the gate pillars. If we knocked off one pillar we'd have more room. Charles laid out the plans for our drive.

We pointed out that the toilet drain ran right under his fine proposal.

'Dig 'er up.'

'Um … is that wise.'

'Not a problem.'

'Ok then.' As we were getting Charles at a brown paper bag rate we decided to do things properly. The mini Pelle was put into action and our front yard looked like a hole.

It didn't take long to find the sewer pipe and when Charles called me from the kitchen where I was watching Sean try to get our chimney pieces apart I was glad of the distraction.

'Did they come like this?' Sean asked to my retreating figure. I pretended I didn't hear.

Charles stood like a man who has solved the world's problems.

'There's ya problem.'

Boomie and I looked at a stupid box arrangement in the ground.

'They all do this sort of shit.'

Shit is the right word.

The sewer pipe from the bathroom comes down a slope, a too steep slope into a collection box. There the 'product' sits until the box fills and it overflows to the other pipe on the right angle to the first and down to the municipal drain. It was never going to be efficient. It was never going to work.

'What do you want to do?' Charles stood over the box.

'Get rid of it.' Boomie and I knew what was coming next. The lid needed to come off. Just another person to see our Australian poo. I felt like taking out an ad or selling tickets.

'Thank God we didn't have corn for tea,' I whispered to Boomie.

The box was full of roots. The pipes going into the box was full of holes where our neighbour the year before had tried to unblock our drain. The other pipe was cracked where Hervè had taken out his revenge. It was a wonder we got 22 days. Charles said he could put proper pipe in with a Y junction and all glued together. It sounded like a fine idea.

'Just add it to the bill.'

Meanwhile Sean had wrestled our chimney into place and it was all looking good. He gave me pointers on how to season the stove, what wood was best and started to pack up his tools. The stove stood in the kitchen like it belonged there. It would keep us warm and dry out the house. I just hoped it didn't fall through the floor.

The bathroom tiles were complete and the new hand basin hooked up. We had to install the shower screen ourselves, but it was a vast improvement on the dank, dingy shower stall we had with a handyman tile job that might have been done by a blind man.

The day of the concrete pour was cold, but luckily no rain. Charles roped in his wife for the spreading about and at 7 am they began to pump. When it was all over, we looked at the level. It just didn't look like we could get the van under the concrete beam.

'Just wait, the eye can be deceiving,' Boomie cautioned. I had my doubts.

When Charles had finished, the bill presented and everything 'Owwright,' he invited us to his house for a drink.

What a house. We heard how they had renovated, removed, retiled, reroofed and refurbished. I wish I had a house like that, but then I thought of the upkeep, the rooms to clean and decided I liked my 4.5 room house with just enough space for two. Sometimes just enough is good enough.

The electricians had put our earth in while the ground was being dug and now they could come back when the concrete floor was set to do the house wiring.

In France you can wire your own house and just get an expert to sign off on all your work. We weren't that skilled in electrics, although we had re-wired the boat, that was 12 volt and couldn't kill you.

Our house was suffering from woefully inadequate wiring. It was that old stuff with cotton and cloth outer. It looked like it had never been touched since the house had been built and most of the runs didn't have an earth, which could potentially kill us.

Our electrician, Steve had a health issue, so things were put on hold for a month or more. That suited our finances, and if we turned on the lights with a wooden spoon we might survive until then.

That first night with our new wood heater installed we set the fire and just sat in front of it marvelling at the warmth, the homely feel and rushing outside to see the smoke going through the grate thingy. And we started to think about wood.

You can get cubic metres of wood, 1.5 cubic metres, trailer loads, seasoned, wait a year wood and the choice is a wide as a Dulux colour chart. People are cagey when it comes to their preferred supplier of wood. If you ask, they skirt around the question, tell you they got it from a farmer friend, a friend of a friend or their father's cousin. No-one wants to let you into the secret of where they got it so cheap. And they you buy off the shelf and they scoff.

'Never! You paid how much. Robbery.'
'So tell us where you get yours?'
'Ah. My father has a cousin.'

We opted for a delivery on our new driveway from the same people who do our diesel heating fuel. I conducted a phone call in French and hoped I got the right thing at the right price.

Charles had said to put a tarpaulin down so you don't get all the chippings in the new white gravel which sounded like a good idea. While we waited for our delivery we started to chop up the furniture to keep warm. Boomie had bought a small hatchet, and give a man a hatchet and he becomes Wild Bill Hickok, itchin' to chop stuff up. It sounds drastic, but we had inherited some old, wood wormy chairs, some old drawers full of holes and other rubbish. This all went in the cleansing fire. We were using our new stove like an incinerator, and for two pyromaniacs, burning anything that looked like it would burn was our nightly entertainment. I briefly wondered what it would do to our new chimney, but only briefly.

When Boomie wanted to chop up the old singer sewing machine, I put my foot down. It was too good to burn. It gave me an idea to sell it on the local buy , swap and sell. We snagged a buyer right away - all she wanted was delivery. We had a van. We could deliver.

Denise lived in a rambling farm house about 40 minutes away. We were invited for a cup of tea and chat after we'd brought the machine inside and once again, made a friend.

Her knowledge of France was vast, her hospitality warming and comfortable. Her driveway was another matter altogether. We slipped, we slid on the wet mud, but couldn't advance up the small slope to the road to go home. Wood was employed under the wheels - that didn't work. It was only by inching forward and trying for tufts of grass we managed to get to solid ground.

'Next time come when it's not raining.'

What day would that be in Brittany? It rained nearly every day. The rain in Brittany is a standing joke amongst the locals.

The wood I ordered came on a pallet, wrapped in clingfilm and the delivery man had an ingenious little electric, all terrain trolley that trundled it over to our shed door. He then took our credit card and left us to it.

The cement had been curing nicely and was walkable, so we loaded the wood in our newly made wood box from our old warped doors, and calculated how long it might last, what the price would be per day and how we might stretch the bottom line. It might have, in hindsight, been easier to calculate how many quacks a duck makes in the first week of May.

When you have wood, you use it.

'I'm putting the fire on,' was heard quite a lot.

What we *didn't* plan for was the slow drying of the house. Our little home had been cold for years and now we were heating it up. There was some serious shifting going on. In the middle of the night we'd shoot out of bed as the floor gave an almighty crack. The windows were sticking and there was a general shift in the ceiling. It was as if the place was coming alive. And we had bugs as they warmed up and came out to play.

What we *could* plan on was our next trip.

Le Mans was on. We had camping gear. What was stopping us. Nothing apparently.

Ten

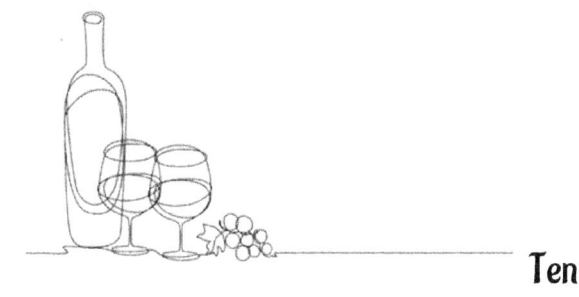

Well, there was one thing stopping us. We didn't have our little van back. I telephoned Michael and politely asked the ETA. I had to jog his memory and then he said to wait. I didn't know if he meant, wait, or wait a moment I will let you know. I hung on the phone for a good ten minutes and was getting ready to hang up when he came back and said we could have it.

Now we could go to Le Mans. I began to pack.

Our camping gear fitted neatly into a big plastic, indestructible tool box, which fitted neatly in the van. We had our portable fridge which could plug into a socket in the back of the van and our bedding covered it all with room to spare. I froze my precooked meals and loaded up on cheese, crackers - all the good stuff. We bought wine from the supermarket, put on our Le Mans t-shirts and were ready to rumble.

288km later we were in traffic looking for the signs to the camping ground. That was one thing the organisers did really well. We were RED camping and followed the stream of bikes, camper trailers, cars, trucks, vans and busses. As we got closer there was a big traffic queue to the gate. At the back you can only guess what is going on up the front. When you get closer the hold-up was security. They were searching every vehicle for guns and things that were out of place in polite society. We sailed through past a van which was being pulled apart, the occupants standing around with their hands in their pockets.

What surprised us were trailers stacked with bonfire wood, old settees, old bikes, engines and even a wardrobe. What the hell was going to happen for three days?

We'd practiced putting up our tent in the back yard so there were no surprises and as we were early we had the pick of the big agricultural field. I pointed to some trees, near enough to the toilets, and far enough from the toilets. It's a fine line. As we were setting up a man came up and asked us to move. He was saving the spot for his mates who have this spot every year. We moved a little to the left and parked the van in front of our door then sat down with a cold beer and watched the constant stream of new arrivals.

And, at the end of the day as the place filled, the fun began. The French began to let their hair down.

Bonfires were fed with pallets and all the wood people had brought. I mused at the safety aspect. We all had nylon tents that would melt at the first whiff of smoke. I could see it was a recipe for disaster, but no-one else gave a hoot as the fires leapt ever higher.

Then we discovered why the old bikes.

Ratbaggery comes in all shapes and sizes. What these motorbike fans had in mind was to load the bikes up with fuel, rev the tits off them and fill the exhaust with oil. Inevitably it catches fire and the result is a flame about 2 metres long shooting out the back of the bike. Some bright sparks had welded huge megaphone exhausts onto their old bikes and the flame was enough to make the crowd jump back with singed eyebrows. Some lads were better than others and had the oil to fuel ratio just right. The noise was terrific as they revved the engine until it popped on the limiter and then hit the stop button making it flame. All good fun.

Some people had gone one better and had big standing engines. Something that might power a truck, bolted down to a pallet and were doing the same thing, but from a distance as the engine was liable to fly apart on rocket fuel. All this 'fun' was watched with booze. A heady combination. Walking around the paddock was great, and as Boomie had his Akubra hat we were spotted as Aussies and welcomed so many places and by so many people.

'G'day,' someone said whipping Boomie's hat off his head. We got it back after we watched a bloke pretend to be a kangaroo. It seems the image of Australia is a kangaroo and Crocodile Dundee and not much else. Some people just couldn't get their heads around Australians coming to Le Mans, like we had made the trip overnight and planned to go home afterwards. One man thought Australia was somewhere near China. But everyone thought a kangaroo hilarious.

The noise of the engines cut through the night, all night. We went for a walk around midnight and things were in full swing. Bonfires, engines, drinking, eating, motorbikes racing through the camp site and burn-outs up against a brick wall. There was noise, smoke, burning rubber, BBQs all mixed with alcohol. It didn't look like the people had any notion of going to bed. We retired to our tent and tried to sleep in fits and starts. I think I managed about 2 hours when someone started up with music at full volume. I call it music, but it was electric techno thing with a beat and not much else.

'We must be getting old,' Boomie said as we watched the red flares shoot into the sky.

We had never seen the French so boisterous. They are, on average, a quite serious crowd. They always do the right thing, they are very polite and hate the thought of making a spectacle of yourself. If someone is drunk they stare like it is the social faux pas of the year. So, to see them let rip, as we say in Australia, was a new experience.

In the morning we spied flags on poles from all the European countries and some English, Scottish, Irish, American and one Australian.

Through the night we hadn't seen one fight or drunken brawl. It was all in good humour.

There was a constant stream of people to the toilets which were everywhere. We waited until the cleaners had been then nipped in for a shower and brush up. I found out the cleaners were coming around every hour and although hundreds of people were using the facilities they were always in good order with plenty of paper. Quite quickly they became unisex as the wait for the men's was twice as long as the females, something you don't see every day. I would have thought the ratio male to female in the camping area about 30 to 1.

Saturday was race day and the track was packed. Being old hands at the race, we were in good time to get a front row view of the start. We knew the drill and I managed to video the run this time.

It's was great being able to come and go to the track. This time we could watch it at night, which is something different. After an hour we headed back to our tent for a rest, there was no hope of napping, so that we would be able to stay up late. Once upon a time in our younger days, we could have pulled all-nighters and not blinked. Now, we needed a rest if we were to stay out. I was reading, sitting in the sun when our neighbours came over to introduce themselves and invite us to drinks in an hour.

They had been watching us and wondering.

I made some sushi rolls, a good easy staple I'd perfected living on the boat. A tin of tuna, an egg, some onion and rice and bingo - all rolled into some nori (seaweed). It looks special, but it's easy and a crowd pleaser.

We trotted over with wine and sushi and made friends for life.

Jerome introduced us to his cousins, his nephew and his work mates. They came to Le Mans every years to get away from the wives, have some boy time and let their hair down. It was a thing. Cecil, his 16 year old nephew was with the men for the first time and loving it.

By the look of it they had started without us and were well on their way to being tanked. And they wanted to know all about us. They didn't speak a word of English and we only have a smattering of French.

With Google translate we told them we liked bikes, we bought a house in France and we previously lived on a boat.

We drank all our wine, then they brought out a 10 litre box of wine. Where do you even buy a 10lt box of wine?

Then Boomie was passed a drink that was a bilious green. It looked like radiator liquid with a glow in the dark sheen. Our camping lights didn't do it justice. They all watched as Boomie drank it in one.

'What is it?'

'Toothpaste,' he said. It was called GET 21 and as far as Boomie could tell it was mouthwash. Jerome and the others loved it. It came in 2lt bottles for around 10€. You can see the attraction.

Jerome was a builder, he said and worked with bricks.

'Ah, Maçonnerie,' I said trying out my French. Apparently it is all in the way you say it.

Masonnerie as in mass- onnerie is a mason. Un con is an idiot. Jerome and his mates heard un con and fell about laughing. How was I to know.

The French had hand signals to go with the conversation, which sometimes helps. A twist of your nose with your fist means drunk. A pinch of the nose means you are a clown. I was apparently both.

As Boomie washed his mouth out once more with GET 21 we were asked what music we liked. We began to rattle off some Australian bands they might know. AC/DC was the only one they had heard of and it was found on their phones and blue toothed to the speaker. Of course, everyone was Angus jumping up and down. We'd heard there was a concert later in the evening on the other side of the track with an AC/DC cover band. That sounded like fun. We went through quite a bit of music and GET 21,then quickly came to the realisation that these men had no idea about any of the bands we knew, and we had even less idea of the bands they raved about. A lot of their music was folksy with drums and anguished wailing. A song would come on and they'd all jump up and sing.

'You don't know?' 'Tu ne le sais pas?' Jerome frowned at me.

'Non.'

He just couldn't believe I didn't know the song.

While all this was going on the engines were revving and someone had come with a chainsaw engine. He was cavorting about the place in a gorilla suit revving the chainsaw, minus the chain until it died. Then he threw it on the fire to the great delight of the spectators. Things were hotting up to a wild night.

Naturally the conversation got around to Crocodile Dundee and his famous line about a knife. And Boomie said when we lived in Far North Queensland we saw crocs all the time, which we did. They would cruise past our boat, our dinghy and sunbake just metres away from our mooring lines. This really impressed the lads. Then Boomie showed them the key ring for the little van. A crocodile foot that we'd bought at the Port Douglas markets. This was handed around, sniffed, played with, and studied with due care. Then, somewhere along the way, Google translate let us down and the Ashwin's were crocodile hunters who had travelled the world on their boat. Our standing in the neighbourhood went from mildly interesting to superheroes.

As much as we tried to explain, it wasn't working, and the first rule of holes is to stop digging. We filled our glasses and were toasted as the most daring people our French friends had ever met. They shook our hands. I was kissed on both cheeks a dozen times. We swapped addresses and telephone numbers and they friended us on Facebook.

We left them drinking to go and see the concert.

'We can never see those people again,' Boomie said as our derring-do began to percolate across the camp site.

How do you top crocodile hunters who sail the world?

Damn you Google translate.

When we arrived at the concert it was just beginning with the warm up act. This would have been hard to take if we were sober. Drunk it was a hoot. About 20 men in kilts from Brittany were jumping around on the stage playing bagpipes and tin whistles. The crowd were loving it. The band was quite famous and everyone knew the words to the songs. I'm not a great fan of bagpipes, I just can't find the melody. The crowd had no trouble. We watched for a bit and then the GET 21 and Boomie decided they didn't like one another. We retired to a bit of grass and then went back to the track for a sit down and a bottle of water.

Watching the bikes at night you know the riders must have balls of steel. Although they probably know the track like the back of their hand, it would be hard to read the road with only the headlight for guidance. Some of the track is well lit, but other bits are surprisingly dark and they are going well over 200km/h.

We sat and watched until some of the alcohol had settled and then went back to the camping area. It was mayhem. Fires were everywhere. The police were just standing around chatting with the cleaners and the hot dog/hamburger stand was doing a roaring trade. Even in the chaos that was Saturday night the French, once they got their take-away would stand around the stall eating. They just couldn't bring themselves to eat and walk at the same time. We bought some *churros and stood about like the natives. *churros are a batter extruded into long fingers and deep fried, then dusted with sugar. If done right they are delicious.

We strolled around the campsite marvelling at the flame throwers, the noise, the doubling of the population in just a few hours. Some hardy souls were pitching their tents on the dirt tracks that were roadways, hammering nails into the hard ground with blocks of wood. It was all so crazy.

After so little sleep the night before, and a gut full of drink I think I passed out instantly. I woke up a couple of times, once when I heard someone having a piss right outside our door, but the next thing I knew the sun was shining and at 5:45 there was quiet. It lasted until 5:47 when a lone engine cracked into life and went to the limiter with a pop. It was answered from far away with another pop and it wasn't long before the dawn chorus began.

And at the showers there was a line up. Boomie was in first and as he was coming out all clean and tidy he was accosted by a drunk fellow, stark naked and full of mud. There was some early morning mud wrestling going on somewhere. He gave my Boomie a big hug, so Boomie turned around and went straight back into the shower for a second go. Coming out again, yet another mud wrestler tried to grab his moustache, but you need to be quick to get him a second time.

There was a bloke asleep on one of the women's toilets and nobody seemed to mind. My shower was incident free, but it was choking on more hair than a barber's floor. There are some things that make me gag. Hair in the drain is one of them, phlegm is another. I had the quickest three minute shower with my plastic crocs on trying not to look at the drain.

Sunday afternoon was the last hours of the race and people were already packing up to go home. To some of the campers the race was secondary to the fun to be had in the paddock. I could see why.

We saw a couple of burnt out motorbikes on a funeral pyre. We saw a few fibreglass tent poles, the only evidence there was a nylon tent attached at some point in the weekend, a small melted blob. We saw little pop-up tents abandoned all over the site. They are about 29€ to buy so just a throw-away. What we didn't see or hear of was one fight, one theft, one accident. One man had fallen in a fire, but was pulled out before any serious damage. I asked a nurse at the first aid had she been busy.

She shook her head. Someone had got their hair stuck in a sleeping bag zip. A couple of cuts.

We headed to the track for the final hours and a good spot to watch the action.

Jerome and his friends were packing up early as they had to get home to normal life … until the next year. We promised to send them a crocodile foot key ring and Cecil a shark tooth necklace, which only added to our kudos.

'I have much respect,' Jerome said thumping his chest.

Oh, my giddy aunt. What had we done.

Boomie and I hadn't had such a wild weekend in years and we made a promise to go back again. We also made a promise to Jerome to visit him at home in Fougères because, he said, he made the best escargot in all of France.

'I don't eat bugs and I don't eat snails,' Boomie said.

'No, you just kill crocodiles for a living.'

'We can never see those people again.'

Eleven

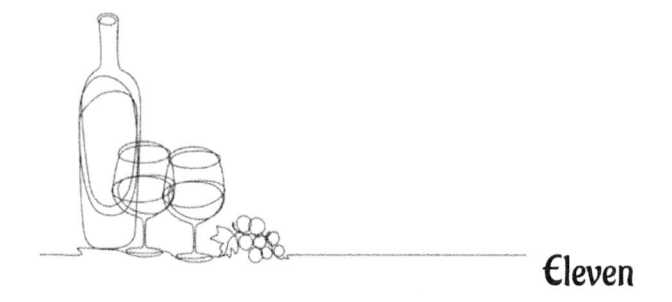

Boomie and I are history buffs. We like the story that goes with the building, the personality that built the bridge and the legend that goes with the terrain.

I am a great fan of architecture. Boomie is a great fan of Cathedrals. So, we decided to do a Gothic Cathedral tour on the bike. We'd only just arrived home from Le Mans, but that didn't stop the Ashwins. We had the travel bug, and bad.

Ages ago we bought a book over the internet on Gothic Cathedrals of France and this had sustained our interest. We actually got two copies as our first purchase was delayed, we complained - well I complained, and they sent another. This book gives the history of the cathedral, what to look for if you stand on this spot, where the stone came from and the small incidentals that are often overlooked. With the book at my side, a map, a credit card and the internet we planned our route across France. Our first big bike trip would end with another 1,200 kilometres on the clock.

I checked the weather. The long range forecast was looking good. We'd be away for two weeks (ish).

I love to do things on the fly and don't particularly like being on a schedule, because it makes you rush from point

A to point B because you have booked. But, it isn't that easy to get accommodation on the hop and so, I booked our trip giving us plenty of time in each spot.

This was going to be the trip of a lifetime. And cathedrals are one of those things that are free. Bonus!

The luggage on a motorbike is minimal. I put out my wardrobe for the trip on the bed and Boomie surveyed my choices.

'Do you need two of those?' I'd tried to colour co-ordinate, and squirmed as I put something back in the drawer.

The panniers on the bike can hold a full face helmet. That's about the size of it clothes wise. If I could wear a lot of thin layers I could rotate them and look fresh. We would need a laundromat somewhere in two weeks travel which halved my wardrobe yet again.

'Don't you have a thinner jumper?'

At this rate I would end up with a toothbrush and a pair of knickers.

It's not the size of the item, but the weight. The bike weighs in at 280kg, then there are two well fed Australians and our clothes. Boomie needs to be able to manage all that weight in tight spots. I tried to tell myself that I could live in the same clothes for a week, no problem. It was that or get left at home.

We packed our things in rubbish bags, just in case and hoped all the hotels I had booked had towels, shampoo and soap.

And on the first day it began to rain. No-one likes to ride in the rain, plus it can be dangerous.

I had planned our first stop in Rouen which was 391km. The map said a good 4 hours. With toilet stops and tourist stops we could do it in an easy 5.

The motorways allow 130km/h, but on a new bike we didn't want to travel that fast, a steady 115km/h might do it.

The motorways in Brittany are free thanks to a deal and probably a handshake made by the King of Brittany and the King of France. Free roads to this day. After Caen the motorways have tolls, but for a Class 5 (motorbike) they are relatively cheap. You could scoot around all the major arterial motorways, but why would you, because the toll you pay is for maintenance, and these roads are immaculate. They also have great roadhouses with restaurants, lovely toilets (always a priority) shops and snack food. They are also marked on maps and signposted so you can plan your stops.

As the rain eased, we decided to make a break for it. Our village is in a rain catchment because of our hill so we kinda knew once we were on the road things might not be so bad. And they weren't.

The sun came out to brighten our day as we rode north to Rouen. I had rehearsed our toll procedure and it all went like clockwork. By the third toll we were experts. I have a little pocket in my motorcycle jacket and with a bit of dexterity I can slip the toll ticket and credit card into a safe place. The French, I have encountered are not very good at waiting at toll booths. The merest delay and they are on their horns. We have had our fair share of tooting and handwaving when we were driving. Now, on a motorbike they just wait patiently. I didn't think we looked intimidating, but the driver's weren't taking any chances.

'Wait 'till I show them my tattoo,' I said to Boomie, 'that will frighten them.'

We don't have a GPS on the bike, but our smart phone does the job. I'd made a cheat sheet which I put in a pouch on my arm and with our two way helmet radio system I can forewarn Boomie. 'At the next exit.' It works well and by 5pm we were in our hotel room feeling quite proud of ourselves. We had negotiated half way across France on our bike, found our Ibis Hotel at Petit Quevilly with only one U-turn and our booking was accepted by the automatic machine.

I love it when a plan comes together.

Rouen in Normandy, is on the Seine River as it winds its way to the sea at Le Havre. It is 135km North West of Paris and has a long history. Joan of Arc was burned in Rouen in 1431, there have been Vikings, Normans, English and somewhere there would be a plaque saying William the Conqueror slept here.

We had planned one day in the city to see the Cathedral, and the old town. This was a Cathedral tour, and we figured we could always come back for more as there is always something more to see, but first we went in search of food.

Ibis Hotels have a microwave for those on the cheap deal. We found a supermarket and bought three minute meals cursing that once again we had forgotten our forks. As luck would have it Boomie's meal came with a snap together fork so he ate first. You just don't seem to be able to buy one fork. 100? Not a problem. One? Nowhere is sight. We took a stroll down the street as twilight crept up and I rubber-necked into people's homes. Some of the town houses had chandeliers, sweeping staircases in lobbies, ornate ceilings and no expense spared. If they don't pull the curtains, perhaps they want you to admire. One can dream, can't one.

We found a tram stop that would take us into the city in the morning, and headed back to crank up the heating, put the television on and eat ice-creams. Sometimes life just throws the good times at you and all you can do is catch it with both hands.

When you travel around a country you live in, it brings a level of familiarity that takes the stress out of things. In foreign places I'm always checking the map, making sure I know where we are, how to get home, what the deal is with buying tickets etc. Boomie just goes with the flow.

'Which way?'

'This way,' I say and he follows.

But knowing how the system works gives me time to enjoy myself that little bit more. And knowing a bit of the language too.

We decided to walk into the city and catch a tram home when we were foot weary.

We walked over the River and into the old city which is dominated by the Cathedral. I had brought the book, and boned up on what we were about to see.

Nothing can prepare you for the sight of a giant set amid houses and bustling streets. The houses around the Cathedral were medieval in character and straight away it takes you back to the 1300s. We looked up to the spire which is the tallest in France at 151mt.

How do people live in proximity to such great monuments and go about their daily lives? Coming from a relatively young country (architecturally speaking) as Australia, we just don' t see what man has achieved over the years like you do in Europe and other places. I guess people get quite accustomed to such things and don't give them a second thought. I would stand about with my mouth open in awe, my imagination working overtime. How many people had trodden this path? How many prayers had been offered up in this place? What was it like when Joan was brought into the city for trial? It all crowds my mind. It's history, a living history to visit these places.

Once inside, it didn't disappoint. The building of the Cathedral had been going on for three centuries so took in all the Gothic trends. After three hundred years I guess just about everyone has left their mark. The stonework is festooned with carvings.

The church was remarkably quiet as we strolled around craning our necks and then looking down at the floor tiles. It really was quite a feat of stonework. I read most of the windows were blown out in the second World War, but in 1939 they took out the stained glass ones and kept them safe. Some of them date to the 13th century.

There are quite a few burials in the church and I saw that Richard the Lion Heart, (1157-199) King of England's heart is entombed in the Cathedral. Apparently his entrails are elsewhere and his body is somewhere else. We had seen fingers, teeth, femurs, and toes of Saints around the place, but to be split into three different sections … I guess he always liked to travel!

Cathedrals follow a tried and tested design, but within the strictures of the transept-nave design the architects can go to town on the dimensions. Rouen's *La cathédrale primatiale Notre-Dame de l'Assomption de Rouen,* is third in the charts of length overall. Reims and Amiens are the first and second place holders. The nave soars four stories high and I can just picture the natives of the city looking up at such a monument and thinking God lives there.

We gawked at the majesty of the place and sat down to take it all in, if such a thing is possible. It makes you feel puny. This was our first Cathedral. We didn't think we could ever get over it.

Across the forecourt is the tourist information centre. It is housed in the oldest surviving renaissance building, the finance office. Straight out of the 15th century. We gathered maps and free pamphlets and went walking.

One of our favourite things to do is wander around the place. We're not interested in shops, so much as buildings, walkways, little alleyways, city wells, old graffiti and how people shaped their cities. We were strolling about when a woman washing her windows nodded and smiled.

'Bonjour,' I said to the woman. Tiny glass panes that had been cast before sheet glass was a thing. They all had waves and bubbles.

'How old are they Madame?' I asked.

'1500 I think,' she said. 'You want to see?'

How could we refuse. We followed her into her house and out to a small courtyard. It had a well, cobbled stone paving and overhanging rooms, all with those tiny windows.

'In my family for many years,' she said in English.

We explained we were from Australia. She said she had a niece working in Australia.

I love this sort of encounter. People are friendly and it makes the trip all the more memorable.

We left with thanks and went in search of a sandwich to eat by the Seine River.

The River was full of traffic, and the blockwork at the water's edge was strewn with seating, parks, playgrounds, neatly kept bollards and ironwork. We sat down with hot chocolate and sandwiches to watch the world go by.

The thing about a motorbike is that we needed to watch our weight. I'm a great pamphlet collector. I would need to get creative if I was going to indulge my fancy. Boomie need never know I had stuffed a map down my shirt front and a free postcard in my back pocket.

'What's that?'

'Nothing dear.'

When you are foot weary an easy tram ride is just the thing. We caught the tram to anywhere and sat down to watch the city pass by. There was no hurry, we didn't know where we would end up, but we knew we could always catch a tram back to the city, so we relaxed and put our feet up for an hour. The tram took us to the suburbs, via the old wharf areas and ended at towering apartment blocks with about a million kids playing outside. We hopped off, and waited on the platform to catch the tram back.

Once more in the city, hunting a glass of wine or beer was as easy as asking a group of young men where the best drink would be. They walked us to a bar and waved us in.

Sitting around with a glass of red watching people is as easy as falling off a log. We had no trouble ordering some snacks, and doing what the French do at that time of day.

'What are the poor people doing?' Boomie asked.

'Who cares!'

Our next stop would be an easy 80km East to Beauvais.

80km meant that we didn't need to get up super early, which meant we could go somewhere for breakfast. There is nothing like a good breakfast to set you up for the day. Coffee and crepes sounded just the ticket.

Back at the Hotel I read about Beauvais Cathedral while Boomie watched a French game show with a lot of shouting. Television is rotten the world over.

Twelve

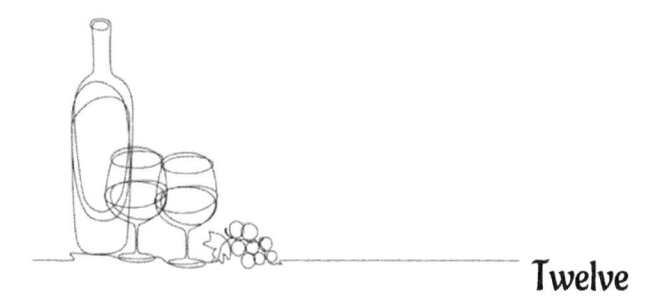

Getting out of Petit Quevilly was super easy as it sits right on the ring road of Rouen. I had my cheat sheet strapped to my arm and as Beauvais was only 80 km away it was signposted. We'd eaten our crepes at a little bar/tabac/take-away and were all set.

Getting into the stream of traffic is always tricky. At rush hour it is near impossible. We had a very near miss with a truck which didn't see us. I think he saw the hand signal Boomie gave him as a parting gift though. A brand new bike might have been totalled if not for Boomie's quick thinking. We are motorbike riders from way back and so are ever vigilant to the stupidity of cars and trucks. Once out of town things slowed down a bit and we could relax and enjoy the ride.

I had planned our road would take us through Fleury-la Forêt. It was touted to be beautiful and in fact, the French said, one of *the* most beautiful villages in France. That was worth a look because we didn't need to be anywhere in a hurry. Usually you can't get into your hotel until 11am, sometimes midday.

Spring in France is a magical time of year. Late spring is a feast for the eye. Communes go all out with 'potted colour' as my mother used to say. They have hanging baskets overflowing with flowers, garden beds on roundabouts and boxes from windowsills, walls and a floral clock might not have been out of place. When a village is declared to be beautiful we expected flowers. The commune of Fleury-la Forêt didn't disappoint.

There was a chateau, lovely little streets with medieval buildings and the sun came out to make it extra special. I hopped off the bike and found a suitable park. (here I was again walking around France with a motorbike helmet on. There must be an entry in the Guinness book of records for that sort of thing.) We found a café, then sat around looking at the scenery and drinking coffee like a local. Sometimes you just need to do that. Just sit and take it all in with your eyes.

We carried on to Beauvais and as we were early for our accommodation went straight to the Cathedral.

This cathedral of Saint-Pierre has a history that is unique.

It was completed in 1272 and then some of the vaults collapsed in 1284. The builder had tried to make something magnificent with the vaults a colossal 159ft high, but the limitations of physics outdid his aspirations.

The partial collapse made the architect have a rethink and they began rebuilding which finished 40 years later. Then, in 1569 the wooden steeple was completed and boasted to be the tallest structure in the world, but four years later that collapsed, luckily, after worshippers had just left the building.

Beauvais was nearly destroyed by a bomb in 1940. The cathedral suffered a direct hit, but only the organ was damaged.

Perhaps you'd think someone might say,

'Let's call it a day fellas.' And yet here it stands over 700 years later for us to admire. That's stickability for you.

Now when you visit, you can see great bolsters of wood, put there in the 1990s, shoring up the walls. It's worth noting that Beauvais Cathedral remains incomplete after nearly eight centuries: The nave, a fixture of most cathedrals, was never built.

There is also an astronomical clock.

'Not another astronomical clock.' I've seen these clock all over the place.

Lombardy, Italy.
Strasborg, France.
Lund, Sweden.
Gdańsk, Poland.
Bern, Switzerland.

Exeter, U.K. and one in Czech Republic. I think I've seen enough astronomical clock to last the rest of my days.

There was a little crowd of people sitting on seats waiting for the chimes. I passed and went to find a quiet space to read my book about the cathedral, and then a cat came up and sat on my lap. I guess he had had enough of astronomical clocks too.

The sun came out and the stained glass put on a show for us. The whole outer aisle was bathed in a red and blue glow slashed now and again with yellow and purple. It really was magnificent. Once again my imagination ran riot thinking of the illiterate people looking at the windows, the light, the height, the majesty of the building and putting their faith at the front of their everyday lives. Frightening and magnificent all at the same time.

The Bishop at the commencement of building, had grand plans and taxed the people relentlessly to see his cathedral built. He taxed them right into a riot until he was taken down a peg or two. You can only push the people so far. France is known for the people pushing back.

We left the cathedral interior and walked around the outside admiring the grand flying buttresses on the chevet and the hefty size of the foundation stones.
'And they did all this with manpower.' It makes you stare and wonder at the capabilities of humans.
Our afternoon was complete with a late lunch in a restaurant on the town square. A lot of cities have a huge open space and they make a lot of sense when you have so many people. Everyone wants to see the sky. In France they do it with a small table between you and your partner, a little biscuit on your saucer and the most delicious coffee. It's civilized. It's very French. A town square is also a good place to riot, guillotine people and organise a protest. It keeps the politicians on their toes.
We lingered over our lunch, watching the people, then went to our hotel to freshen up.

Our evening meal was a microwave job. Three minutes and its done, except Boomie's exploded in the microwave. Beef stew was everywhere and not much left in the bowl. I hunted about but, there was nothing to clean up with except toilet paper and I was just putting the paper in the bin when a couple came into the lobby and saw me with a very stained, wad of toilet paper. They made that, 'I stepped in dog poo' face and hurried up to their room.
What can you say to explain it's not what you think and it's only beef stew - with carrots - and it was an accident. My French isn't *that* good. He ate half of my rabbit and vegetables and we got ready for a night out.

I had heard of the amazing light show at the cathedral after the sun goes down.

We rode to the centre of town, I walked around looking for a park and then trotted off to the *Son et Lumière*.

If you want to see something a bit special, a light show does it. I'd see the one they put on at the Pyramids in Egypt, and one in Paris at Notre Dame and this one in Beauvais was just as good. We were suitably rugged up with our motorbike jackets, scarves and gloves and sat back to enjoy 30 minutes of music, lights, action. When it was over everyone was talking in excited tones as they dispersed across the forecourt.

Beauvais became my favourite cathedral, although it was a hard won first place from Rouen. Our next stop was Amiens.

Thirteen

Amiens Cathedral of Notre-Dame has a special place in Boomie's heart. He has heard about it, read about it, researched it, but never seen it. Now would be his chance to cross it off his bucket list.

We approached Amiens from the south and as you ride on a line of hills above the river Somme, some 5km from the city centre, you can see the Cathedral in the distance thrusting into the sky. We couldn't take our eyes off it as we rode around the city to slip off the ring road to our hotel.

The hotel was called the ANZAC Hotel and situated a three minute walk to the cathedral, two minutes to the centre of town. The Australian and New Zealand connection is very strong in these parts, courtesy of the World War.

The city has a maze of one-way streets and although we could see our hotel, we couldn't get to it.

I got off the bike and walked to the footpath, then with a bit of bravado Boomie mounted the path and followed me to the hotel. There must be an Olympic sport for walking in motorbiking gear … I'm sure of it.

I had specifically asked if there would be bike parking when I booked.

'Oui, not a problem,' the manageress said. There was no parking except on the street, and that was on a meter. I walked around the streets looking for somewhere the bike might be safe and found a lovely man who showed me his parking spot next to his geranium pots. It turned out George was a motorbike man from way back. I rushed back to show Boomie the way, thankfully without my helmet on this time. George was admiring of our bike and said we could park for the three days we would be in Amiens. Some people are just born nice.

Our hotel was once a house, therefore when it was converted to rooms it became a rabbit warren. Add fire doors, staircases over windows, and rooms that just don't make sense behind half a cupboard and you have the whole picture. We were in a half room, you could see the cornices disappearing in the wall, but it was affordable, close to town and the only inconvenience was a shared bathroom and toilet. The lobby was draped in Australian mementos. Flags were pinned to the walls and kangaroos jumped over the reception area.

'Do you get many Australians here?' I asked.

'Yes, very many. We are very good friends here for Australians.' It's nice to be welcomed as friends.

From Beauvais to Amiens is a mere 66km. We had all day to see things. I'd seen some fabulous shops on the way in and wanted to *lèche-vitrine*.

'We can't buy anything remember,' Boomie cautioned me. My pamphlet stack was growing, and I was beginning to look like a Michelin Man. I nodded.

'I know. I know. I'm only looking.' I think my husband had heard those words before.

We took off to see the shops.

Amiens has those wide open spaces too, only it is a huge, wide, paved mall that stretches about four city block or more. It's lined with shops. High end, low end, eateries, drinkeries, and my favourite, book shops. Big book shops in France always have an English section and they usually carry stationary, another favourite of mine. I mean, how much can a fountain pen weigh for heaven's sake.

My new Waterman fountain pen would always remind me of Amiens and it was light, and small, and could just fit in my pocket. Boomie was convinced as to its practicality.

We found a restaurant that looked quite nice for our evening meal and went back to our room to put our feet up for a few hours.

Our chicken stew was delicious, our beer cold and we walked it off in the streets around the cathedral, me rubber necking in houses, and Boomie hoping we weren't lost.

'You know the way right?'

'Yep.'

'Are you sure?'

'Yep.' I have a good sense of direction, always have.

We were woken up the next morning with a very loud crash and bang. A gang of workers were hard at it at 7 am demolishing a building across the road. There was a lot of shouting and not much work that I could see, with the whole thing watched by a dozen other blokes doing nothing. It could have been a scene from any country the world over. People are all the same after all.

The day before I'd stocked up on free maps and pamphlets and after our breakfast in the hotel we headed off to cross the cathedral off Boomie's bucket list.

It's a whopper. It has a special place in the French psyche too, as it is the largest Gothic church in France and on the UNESCO world heritage list. It has the highest nave of any medieval church in the country and more statues on the outside than you can poke a stick at. It was also part of the pilgrims routes in France to Santiago de Compostela. People have been visiting it since it was built in the 13th century.

There are so many things about this cathedral that are noteworthy.

I'd read up on its history and had my book with me to guide me around the interior, but nothing prepares you for the view. I always try and enter a church from the front, that way you get hit with the full effect. The same effect that worshippers would have seen in the 1200s.

It is built in the Rayonnant style, which simply means that they did away with the sturdy, tough, strong look and went for the silken, slender, lighter look for the columns, arches, vaults and stonework. And it works, drawing your eye up and up and up. At 42.3 metres it's enough to give you a crick in the neck.

Aside from the majesty of the building itself, there are old photographs, lithographs and the history of the building all around the edges of the nave, and all of it in French and English.

In the war years the church was sand-bagged right to the roof against bombs. Some of the windows were taken out and hidden for safety as the war raged and some of the sand-bags were to hide the artwork behind, against theft.

Marshal Foch, who served as the Supreme Allied Commander during the First World War, asked the Germans in 1918 to spare the cathedral and not shell it. The Germans said they wouldn't deliberately target the church provided the French didn't use it for military purposes, ie an observation tower. It was hit several times, but the German's kept their word and it survived.

Inside there are memorials to the Australian, Canadian, and New Zealand soldiers who fought in the First World War and as we were visiting just after ANZAC Day there were poppies everywhere.

Boomie knew of a famous statue of 'the weeping angel' in the church and hunted it down. It was put on post cards in the First World War and became a symbol of young men dying in war and was, by all accounts, very popular. The 17th century statue still sits there today, but to our surprise the original was taken away and we only see a plaster-cast today, which is a pity. The original was marble and you would think it impervious to the ravages of time. Now just a trick of the eye.

We ambled around, sat and admired, listened to a bit of the organ tuning up and decided to come back the next day to do the outside and the grounds. You can get a little overloaded with statues.

Right on the dot of 7 am the workers across the road began their labours. As we didn't need to wait for the cathedral to open we opted for an early start.

The outside of a great edifice is as fascinating as the inside. Boomie knew all about the different structures and our book pointed us to the bits of interest.

I have found having a guide book is a great advantage. To know what you are looking at gives a flavour to the experience.

In Egypt, I had a guide book to take me around the pyramids and the archaeological digs and it made the whole thing come to life. 'Oh, so that's why there is a pointy bit on top,' and that sort of thing.

Now we traced our path with our book in hand and marvelled at the workmanship, saw the graffiti and the mason's marks and could put names to the various building phases. The gardens next door to the cathedral and the Bishops palace were equally interesting. There was a little passage for the Bishop to enter the cathedral unseen. Imagine the boss popping up when you least expect him. It would keep you on your toes.

Besides the cathedral, Amiens has floating gardens on the river Somme, called the *Hortillonnages.*

It is 300 hectares in area and has 65km of canals or channels. In earlier times it was all a market garden for the town but today it looks like most of the plots were prime real estate for week-end retreats with fancy houses along-side more modest shacks.

Strolling along the narrow paths you can look into people's back yards and see old water mills, gondola type boats tied up on rickety docks and the elaborate bridges constructed to access the property. The sun was out, people were out and there was a queue for the coffee being served from a tiny restaurant.

From there we strolled around the town and along the river enjoying the sunshine and making plans for the next day.

As luck would have it we found an Irish pub near our hotel and there is nothing like a Guinness with a Guinness pie and pom frites to end a perfect day. We sat with the locals and inevitably the conversation came around to,

'Where are you from?'

'Australia.'

We were treated to a chaser of pastis and a pat on the back.

The French have a great affinity for Australians, the people of Amiens and environs count us as their saviours in some respect because of the war. We stood side by side with the French against the enemy and they appreciated the sacrifice.

We were going to visit the Somme on our trip, and see things that would remind us of that sacrifice.

Fourteen

Boomie has a great interest in the First World War. He has a vast knowledge on the battles, the places, the people. Going to the Somme with my husband is like having a personal guide, which is great to get the whole picture.

We left Amiens early and headed north-west to Albert.

One of the first things you notice as you approach Albert is the Golden spire standing tall above the countryside. This particular morning the sun was striking the orb on top to give us a show.

There is a story about the spire that flew around the battlefield and is the stuff of legends. German artillery shelled the church to stop the French from using it as an observation post, but they only dislodged the gilded statue of the Virgin and child. When the battle of the Somme had ended in 1916 Albert was almost flattened and the statue was hanging from the spire at a precarious angle.

The French and British believed the day the tower was brought down the war would end.

The Germans conversely thought that whoever toppled the statue would lose the war.

It wasn't until 1918 that the Brits gave up Albert to the Germans and then the British big guns reduced the basilica to rubble.

Now we see a replica of the church and the tower, but the legend of the Virgin remains, mainly I think because some enterprising soul back in the day printed postcards of the toppled statue and was selling them like hot cakes.

We rode into the town and I hopped off to find a suitable park … walking around France in motorcycle gear once again.

Straight away we liked Albert. It had a village feel about it and as all the buildings were post WWI, a neat and tidy appearance.

Being too early for the church to open we went to the museum next door.

I would rate this museum as one of the most intimate, informative, genuine museums I have seen, and I've been in a lot over the years. There was none of the interactive, multi-media, open the flap and see something playground about it. It was glass cases, information in English and French, real exhibits and a few dioramas. It was also in tunnels under the church.

A lot of the exhibits looked like they had just come out of the ground, and we saw personal items from soldiers in the trenches. Things like pocket watches, razors, letters, photographs of loved ones and all the objects that bring the people to life. The other thing I liked about the museum was that it included the German side. Many places only concentrate on the Allies, but here we saw German helmets, guns, water bottles, uniforms and you feel sad for the colossal waste of young life who took the brunt of war on both sides.

We had the museum to ourselves so were never rushed and I read voraciously about the objects, the offensives and the winners and losers. We came out like bats, blinking in the sun and had to sit and compose ourselves. After seeing something so devastating, so horribly poignant, we felt drained of emotion. You can read about it, study it, see it in pictures, but when you come face to face with someone's letter home to their mother … it was quite emotional.

The replica basilica was just as interesting as the museum. They had pictures on boards around the church depicting the destruction and the rebuilding. It was smashed to rubble and as we traced the few remaining bits with the pictures as reference, it became, all the more, a remarkable story.

Our next stop after a *boulangerie* (bakery) and the biggest *pain aux raisin* (raisin scroll) we had ever seen or eaten was the Lochnagar Crater.

We rode the Roman road north-east to the village of La Boisselle.

Roman roads are all over France, and some are very easy to recognise because they are so straight. There are quite a few that radiate like wheel spokes from Amiens, and Picardy, the district, is choked with them. The armies that use them in the war, followed the line which was taken by the Legions of Rome. I'd seen roman walls and fortifications in Le Mans as we traipsed around the old part of town. Those Romans sure knew how to make stuff. The walls were part of a house, still being lived in all these years later. The roads were still being used. Sometimes if you look hard enough you can see a chiselled gutter stone here and there and if you are really lucky there are way markers. I love all that sort of stuff.

The Lochnagar crater is, they say, one of the most visited sites on the Western Front. The French have a propensity to always stress the biggest, fastest, best(est) of anything French, almost like they have an inferiority complex, or perhaps they just excel in the grandiose.

The crater is just off the main road and doesn't have a large sign, a big arrow or anything to denote it except the usual little sign you might see on any street corner. When you arrive there is a car park, but no visitor centre, no interactive voice-over or coffee shop. There was a small caravan selling Somme dirt, and spent bullet casings from the Great war, but I was relieved to see the absence of commercialism in this solemn place as there are still the remains of soldiers in the crater. In fact, all through the area soldiers still lie in the ground. The farmers are allowed to till the soil, but if they find anything, work must stop while the remains are examined and then interred at the numerous war cemeteries dotted around the countryside. We heard stories of farmers, hastily reburying things, because the paperwork involved, the time taken, is a pain in the neck, when you are trying to earn a living on the land. Never-the-less, soldiers are still found on a regular basis and reinterred with due ceremony.

In reality, the Lochnagar crater is just a big hole in the ground, albeit the largest crater ever made by man in anger. (there is the French grandstanding again) It is now privately owned by Richard Dunning who bought it in '78. He had read about the crater and set off to find it. Once found, he was so moved by the story and the sight of the explosion, he decided to buy it for preservation. After a few years of negotiations with the French he succeeded and, the story goes, to this day, he has never revealed the price, quoting,

'It was a perfect price for a hole of its size and condition.'

The story of the crater is a fascinating one. Boomie knew all the details and with reading the information boards and listening to him, I got the whole battle scene before my eyes.

The Germans were dug in, and so a tunnel was devised by the British, to place explosives under their position.

The 60,000lbs of ammonal was set in charges 2 metres apart and 16 metres below ground. This was the largest mine of the 17 exploded on the 1st of July 1916.

The explosion obliterated - that's the only word to use - obliterated 122 metres of German dugouts which were supposedly full of German troops.

The debris rose 1,200 metres into the air and the history books say that height is three times the height of the Empire State building.

Seen from the air it is a huge scar in the ground 100 metres across, 21 metres deep and 4.6 metres high. I know I can quote figures, but let's just say it's big, really big.

We walked around the edge and from the vantage point you can see all the markers of the battle that raged over the land. Boomie pointed out the woods where troops sheltered and fell.

If you see cemeteries scattered across the countryside, rather than in one place it is because the French buried the soldiers near to where they fell. What is remarkable about that fact is that all the cemeteries we saw were well kept, neat and tidy. The Germans have graves too, but they do not gravitate to memorials on the actual spot, like the Allies.

When you read about the Great War it all feels like it was eons ago, but when you see the battlefield it reminds you that it was just 100 years, or two generations ago, which isn't long in the scheme of things.

Carrying on the Roman road we rode to Thiepval.

Here you can see the largest Commonwealth war memorial in the world. It's a striking monument and what was remarkable to me is the fact that you can't see it from the road, and you can't see it from the car park. And this time there were plenty of parks, so I didn't need to do my reconnoitre for a suitable spot. (Damn, I would miss out on my gold medal for walking across France, with full motorbike gear on.)

The Thiepval War Memorial is set in the woods, a quiet meditative place, as it should be. Of course, with a big thing like this, there were tour buses, a visitors centre, shop and museum. We'd hit the tourist trail and school visit period.

The monument was designed by Edward Lutyens. He was the darling of the 1920s and everybody wanted Lutyens to 'do' their house, their memorial, their municipal buildings. From India, England and South Africa he put his architectural stamp on things. The Thiepval memorial is Art Deco in theme, and was built from 1928 to 1932. Some hated it, some loved it. I loved it. (but then, I'm an Art Deco fan from way back). It has a sturdy, permanence about it and over its columns are the names of more than 72,000 British and South African forces who died in the Somme region before March 1918. When they find someone else in the fields, to this day, his name, if known, is chiselled in place. Walking around the monument we were lucky it was nearly lunch time and the crowd was thinning. As I've said, everything stops for lunch.

I managed some great photographs with no-one in them, always a bonus when taking holiday snaps. The place is kept immaculate, the visitors are respectful and although there were children they didn't romp over the steps, they spoke in whispers and I felt that the younger generation might just turn out okay after all.

The French schools teach the war with field trips to all the pivotal places. We were to encounter many buses with kids all toting question and answer papers. For the most part they were polite, respectful and studious and not a distraction.

From Thiepval, it is only a short ride to the Ulster Memorial Tower. This is Northern Irelands memorial to the war and the 36th Ulster division whose casualties numbered around 5,000. There is a small café and we stopped for a cup of tea, our money going to the charity that keeps the place open for visitors. The café too had souvenirs from the battle ground, with bullet casings, webbings from uniforms, buttons and the like. The ground is literally covered in the remains of the war. The ladies in the tea room were friendly and wanted to know all about us. I came away with a bookmark, and we left our spare change in the charity box. These older people are a dying race of volunteers. The woman freely admitted there was no-one new to take her place when she called it a day. In years to come, would the tower just be a folly on a hill, the true meaning forgotten with time. It happens like that sometimes.

Our day complete, we headed back to Amiens. I had some serious hiding to do regarding the free pamphlets and maps I had collected ... or I could come clean - tough choice.

The next day we would be off to the cathedral at Noyon, but first we needed to buy George, our geranium pot/ parking man a bottle of wine.

Fifteen

To get to Noyon we went via the Australian War Memorial at Villers-Bretonneux. It was a cold foggy morning and we were on the road by 7:30. It felt a little surreal riding in the fog on another Roman road. This one was amazingly straight, cutting through small hillocks and over the flat plains with a dead eye for accuracy. Those surveyors of old knew their stuff. The road would travel all the way to St Quentin, but we would be turning off to head south.

Villers-Bretonneux has a special place in its heart for the Australians who fought to take it from the Germans. The Germans had tanks.

Now, the locals keep the connection alive with ANZAC parades, a school donated by the people of Victoria Australia, and they have the Australian memorial just out of town.

When we arrived at the site, which again was designed by Lutyens, we were alone. The fog was lingering over the graves and the whole site took on a reverence I have never felt before. It was quite moving to walk quietly down the avenue to the main building shaped like a three sided quadrangle with a tower in the middle of one side. It also made for fantastic photographs.

We felt the weight of war on our shoulders as we walked the aisles looking at the names of the fallen. There were wreaths on a lot of the graves, left over from ANZAC Day and messages from families to remind you that these people had lives besides the war.

The sun was just rising above the tower when we climbed up the steps to the top and watched as the fog slowly burnt away revealing the true scale of the monument. Something I will never forget. Lutyens certainly wasn't a one hit wonder. It really was a magnificent edifice.

Originally the design for the memorial was thrown open as a competition, with only Australian veterans and their parents allowed to enter, and the stone was to be quarried in Australia. But with all good ideas the cost, the depression of '29 and myriad of other problems halted the project until Lutyens came to the rescue. It was completed in 1938 and opened by King George VI of England and in keeping with the modern times his voice was broadcast directly to Australia. With the time difference I wonder if he had much of an audience in the Antipodes.

There is a story that in the Second World War, the Germans used the structure as target practice and you can see scars on the block work, but I don't know for sure it was the Germans. Still, the story makes for good reading.

Noyon seems like a small town to own a big Gothic cathedral. Naturally, in keeping with the biggest, the tallest and the best, the French call it 'the earliest Gothic cathedral'.

It took 85 years to build, which would be a lifetime or two for those of the 1150s, when most people died in their 30s and 40s. The site has seen 5 churches come and go and this latest incarnation is called Romanesque-Gothic. The Romanesque bit of the picture is the rounded edges, the rounded towers and all the other round bits.
It had heavy damage in the First World War and then took another 20 years for the restoration. You can only guess at the price of things just after the war. Before that, the French Revolution took a bite of it, destroying the Bishops tomb and probably robbing the stones in the process. Nothing is sacred when you are hungry, the Government is trying to tax the skin on your back and someone is telling you to eat cake, (that bit is only hearsay).

Noyon as a city has seen its fair share of celebrities pass through. Charlemagne was crowned King of the Western Frankish Kingdom of Neustria in Noyon in 768 and the King of France, Hugh Capet was crowned in Noyon in 987. You don't need to dig too deep to find a plaque suggesting someone famous slept here, died here, or fell off a horse here - I saw a plaque in the UK about such an event, and so unmemorable it was, I've forgotten who fell and where.

We had the whole place to ourselves and it was great to walk around in silence, just looking and reading our guidebook. The tiles on the floor are set in an optical illusion that makes the whole thing look 3D. I love locks and window latches and Noyon had quite a hefty locking arrangement for the front doors, and considering the revolution troubles they had put an iron fastening in place to take a plank of wood about 200cm thick. No-one was getting in, or out when the door was locked and barred.

The cathedral is set in well-kept grounds and it was to this sunny place we retired to sit and look at the gargoyles, some very naughty ones, others of devils eating children and just the usual debauched behaviour. The chevet (that round bit at the back) and the scale of the building in relation to its near neighbours made you realise just how big it was to mere mortals. Little houses crowd up to it, to be swatted back by the majesty of the church.

Right around the back was a medieval building housing a library of the scriptures of old. This was wood and daub construction and had lead windows similar to the ones we'd seen at Rouen. Once again it made me think about the people who see this sort of thing every day on the way to work, for near on 900 years.

We also saw a building, said to be the oldest medieval building in … here we go again with the oldest, the biggest, the most(est). France doesn't need to aggrandise itself. It has everything, everywhere. Its rich history, its magnificent buildings, its chic, its Frenchness speaks for itself. And the food, the wine, is a whole 'nuther level.

We didn't linger in Noyon but carried on south to Compiègne.

If you know anything about history, the name Compiègne is synonymous with the end of the war. It was here, in a railway carriage in the forest of Compiègne that the armistice was signed, ending the Great War. The date was 11-11-18.

There is a clearing where the railway carriage sits and you can do a walk through to see all the memorabilia. The only downer is that the original carriage was burnt or destroyed by the SS when the Second World War was nearly over.

The one we saw was of the same vintage, the same stock as the original and had been rebadged to replicate its original twin. The carriage was also used by Hitler, because of the ignominy of the first signing, when it came time in the Second World War for France to capitulate he had the carriage recalled and made the French sign, in 1940, at the same spot.

The museum had, besides the carriage and all the history, lots of photographs of the war, the era and the major players. These photos were old 3D, a new invention at the time where there are two lenses and you view the picture through a sort of binocular contraption. I tried it and immediately felt sea sick. I couldn't look and I felt like I was going to throw up. Boomie saw it all. I could only look with one eye shut and then you only get half the story. Eventually I had to go outside and get some air. I like to throw up in private and headed for the woods.

The forest stretches for kilometres and in the dappled sunshine it was a glorious ride on the way to our last cathedral at Soissons.

What we didn't see was the huge chateau just half a kilometre away in the village of Compiègne. Oh, well, you can't win 'em all.

The road to Soissons was due east and our hotel was waiting; it had been a long day.

We were staying just out of town in a charming little place with flowers in the window boxes, all the small café tables you could want in the garden and a woman who would bite your head off if you looked at her the wrong way.

She took our names, our credit card and pointed. Not a word about breakfast, where our room might be in the rambling old house or what time the front door shut.

After a lengthy search we found our room and flopped. A nap before the evening meal is sometimes necessary, but we napped well into the evening and then it was too late to go anywhere, so we just went to bed. I don't think I've slept so well for years. All that fresh air does you the world of good.

The grumpy woman excelled at breakfast and redeemed herself with hot croissants and jam, crepes, coffee to die for, and stewed fruit, home made by the look of it. We sat back replete and I read the book to Boomie on Soissons.

This cathedral was almost rubble in the 1914-1918 war.

The Germans bombarded Soissons with just about everything they had and then, as they retreated they gave it a final parting shot and left most of the city devastated. Eighty percent of Soissons was destroyed.

We rode into the city centre aware that probably all the buildings we were seeing, were post WWII.

I hunted for a park and we walked to the cathedral situated amid shops, car parks and the hustle and bustle of life almost like it was put there by accident.

When the war ended all that was left was the façade and some flying buttresses on the north side. If you look you can see the pock marks from the artillery.

Now it has been completely restored, at what cost I dare not think. The French Government must have deep pockets for such is the artistry of the work, you know it wouldn't come cheap.

We were lucky to be the only people in the church for our first impression. It's a beautifully proportioned church. The story goes that the church stained glass has suffered from various quarters. The Huguenots, the Revolution and the early restorers sold pieces to private collectors. So, what we see isn't medieval but nineteenth and twentieth century windows.

Some of the churches we'd seen had comfortable seats, cushions and heaters overhead. This one has those cheap rush woven chairs you see in all the cafes and combined with the cold underfoot it might not have been so comfortable. The pews at Beauvais were as old as the building. The seats in Amiens were old, worn and polished to a high shine from hundreds of bums. These looked like they ran out of money. A juxtaposition of terms for the Catholic church.

Our trip was coming to an end. Soissons was our last pre-booked accommodation. I had left the trip open ended in case we wanted to go somewhere else on the way home. But we needed to get back to renew our visas at the prefecture, fix up the house a bit, and do a bit of gardening. It's all very well to go galivanting across France, but sometime you need to come home, wash your clothes and do a bit of renovating.

We left Soissons the next morning and had to make a choice. We could go through Paris in a straight line west, or we could retrace our steps to Beauvais, Rouen Caen and home.

Paris gave us the willies last time we drove through the streets. Unless you know exactly where you are going, it is hard work. We didn't have a GPS to tell us to turn left, and although it was a quicker route we opted for the safe and familiar. Besides, we knew where the hotel was in Rouen and more importantly how to get there. It was a cool 580 km to home via Paris. 600 ish via Rouen. We could do it easy in two days via Rouen.

Our skill level had shot up as we negotiated slip roads, roundabouts and motorways. We felt at ease on the bike and knowing the route made it all the more relaxing. There was one moment when I was so relaxed I fell asleep and woke up with a jolt hitting my helmet on Boomies. Not a good look to fall off a bike because you fell asleep.

The kilometres ticked off without effort and we found the hotel in Rouen. It was fully booked. I rang another on the other side of town. Booked out. Was there a shoe salesman convention or something? I rang a third. Nope. There was a room at the most expensive place in town, but that didn't feel like an option. Who can afford 300€ a night for clean sheets.

The only thing to do was keep going, hoping for an auberge somewhere along the way. We hit the road heading for Caen and knowing pretty soon we would be back in Brittany.

Caen has an amazing ring road. Once you are on it, it's like a superhighway. I saw Ibis hotels, some had no vacancy signs. Did we miss a national holiday or something? It was later we found out that there is one public holiday every week in May. Just our luck.

With the kilometres ticking off we decided to make a dash for home. If we kept up a steady speed we'd be home by 7ish. It would still be light at that time and the supermarket would be open. Wine!

I can't say which was my favourite cathedral on our trip. Each had a moment of its own in my memory. I couldn't pinpoint what town caught my fancy, although I think Albert was high on the list.

We sat back, with our wood stove keeping us warm, and thought the idea of a themed holiday the best idea. You don't get overly distracted with flitting here and there trying to see it all. You discover things at a more leisurely pace and you get to know the subject without other things clouding your memories. I came clean on all my pamphlets and maps and we revisited the places.

'I knew you had them.'

'Oh, how?'

'Hettie,' my husband gave that look that only someone who has been with someone else for nearly 30 years can give.

I think it was then we decided that next year we would go to Verdun and surrounds and immerse ourselves in the history of the trenches.

Meanwhile we congratulated ourselves on the drains that worked, the cement that had cured and the email from the electrician to say they could start in a day or two.

We just had to get our visa sorted and everything would be tickety boo.

And then we had an email from Boomie's brother. He would be in London in a week for a couple of weeks. Would we like to meet?

London?

I booked our tickets on the Eurostar. I'd never been in the Channel Tunnel aka The Chunnel.

Sixteen

We gave our electricians two weeks, then we would be off to England for a 5 day break. I had lived in England all of the 80s and a bit of the 90s. A lot can happen in thirty years, and I was looking forward to the trip. If I didn't pack too much I could bring back books. You gotta take it when you can get it!

The electricians were ready to get cracking and took to our house with a vengeance and a lump hammer. Want a light switch? Not a problem. Smash a hole through the wall.

Our internal walls were made of cheap thin bricks because just after the war that was all you could get for the price. Some of the walls were built right on the wooden floors. You can see why the floor was sagging in places.

Now the electrician was demolishing the house. We had holes everywhere and every night I lamented our poor little house taking such abuse. How were we ever going to fix a hole which was there for no particular reason except she had made a mistake. The husband and wife team were relentless as they punched their way through to route new cables. They would pull the old ones out with a yank that brought plaster and everything else with it. I went for long walks. I couldn't stand it. Boomie was made of sterner stuff.

I knew they could plaster over the mess, and when it was done and I'd sanded and smoothed and replastered and sanded again you wouldn't know, but I knew. I knew what lay beneath.

In the end we had a new switch board with an RCD and knew when we turned a light on it wasn't Russian roulette as to the probability of surviving electrocution.

We had LED lighting under the house that for the first time illuminated every corner and we had fancy spot lighting over the sink. I could actually see what I was washing up. It all happened over 10 days and that was that.

Next was our visa appointment. So far, so good we arrived in good time and took a seat to wait, and then a woman came out and told everyone that the computers were down.

These things happen. We sat for a bit and then people started to leave. Not sure what to do, or expect, we waited.

Then, miracles of miracles our number was called and we went into her little cubical to be told they couldn't do anything for us.

'But,' I began. She held up her hand and I shut up. All she could do was give us a scrap of paper with a hand written note to say we had come in and the computers were down. We would be advised of a later appointment.

Well that was a bother, because I had booked the tickets to London, no refunds - it's cheaper that way.

I tried to explain that we were going places and the woman just waved her hand in the air and shrugged. I didn't know if our little slip of paper was valid for three days or three months. As it was just on lunch she shooed us out of the office and that, as they say, was that.

It wasn't much, but I photocopied our piece of paper, took a photo of it and put it under lock and key. (not really, but you get the idea.) This was the only thing keeping us from being chucked out of France for being overdue on our visa renewal. I'd read about things like that. It could happen!

I got Myriam to ring the Prefecture and ask if we could refuse the date for renewal if it came while we were away.

The French are sticklers for procedure, but when it all gets too hard, they usually give a Gallic shrug, throw up their hands and raise their eyebrows and things just happen. The woman said we could make an appointment right there on the phone,

'Not a problem.'

I picked a date into the future and asked for an email confirmation. It pinged right back at me. How easy was that. Now, London could be enjoyed without the sword of Damocles hanging over our head. I printed the email on Myriam's machine, photographed it and put it under lock and key. You can never be too careful, or blasé.

As the cement under the house had cured, we could now park the bike and the little van and lock the door. We just needed to get the little van *through* the doors. Tricky.

We had 5mm each side of the sticking out mirrors. If Boomie drove on a slight angle that just might do it, but we didn't have a lot of wriggle room.

He inched forward and I directed the traffic. These things need to be taken slowly, steady as she goes, and with patience. Boomie had three goes and then we went in for a cup of tea.

If we took the by-fold doors off, that might help. But the hinges were old, the wood soft and once off, they might never go back. If only the mirrors would fold.

I went on Google to see if somehow they might snap in towards the body of the car. Wouldn't you know it, ours didn't as far as I could see.

'Let's give it another go.'

He inched forward, I darted from one side to the other watching for that precious 5mm and over time we made it. Getting out would be hard too, but that would be after our trip; no need to get worked up about something in the future.

As I suspected our floor was a tincy wincy bit high. About 5cm too high and the van only went as far as the top of the windscreen then would touch the beam.

'If we take the aerial off,' I suggested. I'd seen in the handbook you could de-mast. With that off we gained enough that we could close the doors. Made it, phewww.

This time we could catch the bus to St Brieuc, I checked and there were no school holidays, public holidays, Saints days or strikes. Our local bus gets in with plenty of time to catch the train to Paris, then we just had to negotiate the underground Metro from our West bound station of Montparnasse to Gare du Nord. Our Eurostar left around 4pm and in two and a half hours we would be in England. Don't you just love that you can be in another country in a couple of hours. The train goes about 300km/h on land and 160km/h in the Chunnel.

Getting around in Paris is easy if you know what you are doing. I had it all sorted, except I didn't plan for rail works on the metro line, which necessitated us taking another line, changing, taking a third, changing back to the original line on a different platform that meant when we finally arrived at Gare du Nord we were foot weary, hungry, thirsty and had only about an hour up our sleeve to go through passport control, baggage control, more passport control and get something to eat. The Ashwins always arrive with time to spare, this was tight for us.

I lined up at the boulangerie and Boomie scoped out where to go. Eating while walking isn't the done thing, but we were in a hurry and then a beggar came up to me and asked me for my half-eaten bun. Now, I'm not adverse to giving someone a helping hand, a leg up, a few Euros, but a half- eaten bun? I hadn't had anything all day and so I said,

'Non.'

He didn't like that one bit and proceeded to swear at me, wave his arms about, make a fuss, and generally gather a small crowd. I wondered if it was a ploy to catch us off guard and pick our pockets. This was the big city after all and we were two, well dressed, older looking people from the country. It can happen like that. Boomie pulled me away from the gathered crowd and we jumped on the escalator like sprinting athletes.

At passport control on the French side we showed the woman our precious bit of paper. She read it and frowned. We smiled. She called her colleague over and she read it. She frowned. We smiled.

We waited.

'I knew it. I knew it would be trouble,' Boomie was beginning to fret. I kept my trap shut and looked at the woman with, what I hoped was a confident stare.

The ladies conferred for a bit and then she picked up her stamp and BINGO we were through with a wave of her hand. It looked like it was all too hard for her. We were leaving France anyway, what was the big deal.

On the English side of things, for they have the leaving and arriving in the same place, we passed our piece of paper over.

'Is this all?'

'Oui, I mean yes.'

'They gave you this?'

'Yes.'

'Now I've seen it all.' The woman rolled her eyes. I guess she had seen all sorts in her job. This was a first. She gave our passports a stamp and we were on the British side. So far, so good. We just had to get home again.

And you wouldn't believe it, when we boarded the Eurostar, someone was sitting in my seat.

We'd always had no trouble with Ibis Hotels and so I'd booked an Ibis Hotel in London, well actually in the outer London district, at a reasonable rate, keeping in mind the proximity to public transport. What I didn't figure on was the location in relation to the less salubrious parts of London, and there are always less salubrious parts of a big city.

We purchased a 5 day underground pass from a machine with unlimited trips, I made sure our 'out of town' hotel was within its limits, and as twilight extends to around 10:30 in the northern hemisphere we were in plenty of time to get to our digs.

It was just as well we arrived in daylight. I wouldn't want to walk the streets at night.

'Where the hell are we?' Boomie said as he picked a bit of rubbish off the wheel of our suitcase.

'I don't think we're in Kansas anymore Toto.'

To get to our Ibis we needed to walk through a park, at twilight, looking like two yokels come to the big city, albeit well-dressed yokels. (even yokels have standards).

I consulted our map and because there was a canal in our way, this was the only way to the bridge, unless we wanted to take our chances at crossing a motorway.

I put my brain on high alert, put my backpack under my coat to look like Quasimodo, Boomie gripped the handle of our worldly goods a little tighter and we made a dash for the other side. I've never been so glad to see a neon sign in my life. We heard a scream and a screech of car tyres as we entered reception and jumped right out of our skins waiting for the automatic door to go click and lock us in.

I didn't want to venture out for a take-away so we survived on a mars bar from the vending machine and a hot chocolate. Vending machines have no idea on how to make a decent hot chocolate. Still, on the up side, we had survived the trip, we hadn't been robbed in Paris or London and we were in England. Land of bookshops and people who speak the same language.

It had been years since I'd been in England and a lot had changed while I wasn't looking. There were twice, if not three times more people and they were everywhere. The tube was packed at all times of the day, the roads were congested more than I remembered and the prices were high.

We were to meet Boomie's brother and his partner for lunch just off Trafalgar Square. All we needed to do was get there for 12 o'clock. I don't know how it happened, but we ended up in the middle of Oxford street at 12:15 trying to hail a taxi. Where is a free taxi when you want one? I saw one on the other side of the street and waved frantically, then we ran and jumped in.

'Havin' me lunch luv.'

'Merde.'

'Where ya want to go?'

I'd lost the piece of paper with the address and could only remember some of it. We rang the brother and he dictated the address.

'We are on our way.'

We hailed another cab and hopped in, and had the ride of our lives. He took us up, down, around, through and finally we arrived. I'm sure there must have been a shorter route, after all, I'd studied the map, knew a bit about London.

'Forget it,' Boomie's voice of reason and calm came through my teeth grinding. I hate to be taken for a ride.

A couple of bottles of wine later we were feeling all was right with the world, London wasn't that manic and as we were near the National Gallery, after lunch we'd stroll around and look at paintings.

After lunch I think everyone in London wanted to look at paintings in the Art Gallery. We took one look at the security measures, vis-à-vis taking our coat off, tipping out your bag, going through a metal detector and some people were being frisked and decided we didn't need to see a Constable, a Monet or a Titan with the rest of the population of London and half of the United Kingdom. What about Westminster Abbey? They wanted over £20 to go inside. After all the beautiful cathedrals we'd seen in France for free it felt like a rip off. We'd both seen inside before, when it was for free years ago so left with our money intact. We went shopping instead with the other half of the United Kingdom who were in Oxford street.

We walked there and it is amazing that it all comes back to you after many years. I kinda knew where I was in relation to things, but when we got to the corner of Oxford Circus I just needed to get my bearings so I asked a woman if this was Oxford street.

'No, you need to go down there,' she pointed to Regent street and walked on.

I didn't quite believe her. I asked a shop assistant.

'Yes, luv, this is Oxford street.'

Why would a woman send us off in the opposite direction? I couldn't understand it. The French would walk you to your destination, probably over hot coals if necessary.

I'd always wanted a trench coat and that was the quest we set ourselves. I found one I liked, at a price I liked and then I found a beret I liked, at a price I liked, and a scarf. After a successful shopping expedition it's good to just relax. What better place to relax than a bookshop.

Books in the U.K are cheaper than in Australia, don't ask me why. We entered a large shop and split up for a good hour. This shop had little nooks where you could sit and read, peruse and flick through that coffee table book you knew you would never buy, but was nice to look at.

I came up to the counter with about four books, Boomie had five. We only had one little suitcase, but that never stopped us from indulging ourselves. With a bit of jiggling I was sure I could fit our stuff in, and at a pinch we could buy a second case. The idea had merit.

In the middle of the night we were woken by sirens, screams, cars racing and flashing lights. Something big was going down in the neighbourhood. The evidence was blood in the park, which we saw on our walk to the tube station.

'That could have been us,' Boomie said.

'Let's get home before the street lights come on.' Sound advice.

With the lunch accomplished the day before, we went sight-seeing. Boomie wanted to see the Cutty Sark. I wanted to see the Maritime Museum and have a boat trip on the Thames.

The Cutty Sark was impressive, with the café under the keel in dry dock, but the experience left us a little flat. It was made for children with flaps you open, interactive things and voice-overs that interrupt you as you read. The shop had more information than the boat.

There was, however, a lovely pub right next door and we enjoyed fish and chips and a pint of ale; the real English experience. Though, I did miss my expresso coffee, just a little.

Onto the Maritime museum and this again was kid-centric. Colouring-in tables, screaming kids racing through the museum, treasure trails and pirate adventures all competed with exhibits. There was none of the polite, respectful and studious about the school children as they raced around collecting stickers. The museum was partly to blame as they had made the whole thing into a theme park. I doubt if the school children came away with anything worth remembering. The only quiet spot was a viewing room of the spectacular painting by Turner of the battle of Trafalgar in 1805. You can get quite close to the colossal painting and it is a masterpiece, painted in 1822. The lighting is fabulous, the room quiet and reflective.

We did see Nelson's jacket with the musket hole. He was quite a small fellow, as were people in that era. There was a good selection of guns and handcuffs, a very interesting exhibition on the slave trade and model boats that would test the patience of an adult sloth.

We had a coffee in the café and went in search of post cards, another thing I like to collect. (I have no shame in collecting things)

After museum walking for a few hours, a boat trip is just the thing. Our tube tickets included the Thames ferry service so we hopped on and spent an hour watching history pass by, one magnificent building and bridge after another. Rivers are the life blood of a big city. The Seine River cruise is worth the money, this Thames boat ride was just as great.

I love public transport. If you have the right ticket you can hop on and off anytime to go anywhere. We alighted at the Naval College and walked around the grounds, me reading from a free pamphlet I'd picked up, when we heard music. Wonderful, full orchestral music. There was a rehearsal going on somewhere, but what a treat to sit on the grass and listen to Bach, Debussy and what Boomie thought was Handel. It's moments like this that make the day perfect.

I'd read about the Mudlarks, those people who look for things in the river mud, usually old things. It was also low tide on the Thames and we saw some mudlarks fossicking so I asked them what they'd found.

Several clay pipes were shown to us, a coin from 1930 and a belt buckle. The father and son did this sort of thing every weekend there was a low tide. There most impressive find was a Roman coin, the lad said he liked the bullet he found. We were offed a clay pipe as a souvenir, but declined knowing how fragile those things are, but came away, once again knowing people are, on the whole, a friendly bunch.

We wandered back to our hotel with 3 minute meals; this time we'd remembered to bring forks.

The last day and we wanted to see the Imperial War Museum.
If the Maritime museum was disappointing, this made up for it. It was quite spectacular. The displays were huge, well-constructed and when you are met with a Spitfire dive bombing the first foyer you just know it's gonna be good.
The thing I liked about the museum was the ability to get up close. You can touch the replica of the atom bomb that exploded over Hiroshima, called Little boy, you can see the eagle and Swastika which stood over the Reichstag and the 15 inch guns from battleships which served in the First World War. There are also quiet spaces where you can read about the exhibits and these little stories give great insights into the ordinary man.
The room that fascinated me was the one dedicated to spying. Here we had code machines, tiny pistols, secret compartments in shoes, and teeth, the ingenuity was boundless. It also brought women into the war, which I liked.
As it was our last day we went all out and had lunch in the museum. People do this sort of thing all the time, but we are more the sort of packed lunch, Marks & Spencer sandwich types. It didn't disappoint and made a great day, better.

We walked out into the parklands adjacent to the museum full of the sights we'd seen and fish and chips when Boomie's zip on his puffer jacket got stuck. It had been giving us trouble for ages, but now it wasn't working it all.

If you see an old couple in a park, she's on her knees in front of him and she looks like she might get arrested, or given 5€, she's only trying to get his zip to work. That's all I was trying to do. I wasn't thinking how it would look to the man in the street, and the man in the street did look. Someone shouted,

'get a room.'

I couldn't help myself'.

'Get your mind out of the gutter,' I shouted back and the zip freed itself.

'As soon as we get back, it's going in the bin,' Boomie said. Sounded like good sense to me.

The only blot on the horizon was getting back into France with a small, 'Hettie forgot her packed lunch today,' note from the Prefecture, because we were over our allotted visa allowance and once it expires … well the wheels of French bureaucracy might turn slow, but cross the road and you will be run down.

At St Pancras train Station the French official didn't believe our small scrap of paper.

'Non.' He handed it back. What now?

I suggested they ring. Then Boomie said to show them our appointment email.

I handed over the photocopy, I opened the email on my smart phone and handed that over. What more did they want? A blood sample? Our mortgage docs? A spot on our Christmas card list? There was a lot of conferring and gradually I could see they were giving up. People drifted away from the huddle, some shrugged, one man pulled out a cigarette, the better to contemplate. It was all too hard.

Then we saw the Gallic shrug.

We were going home.

Seventeen

Summer was just around the corner and after travelling all over, we were looking forward to staying put, doing some small renovations and winding down. It would only be day trips around the district for a while, which is very civilized and costs a lot less money.

I decided to try my hand at some French cooking and bought a book, in French, from the local supermarket. The man on the cover looked like he knew what he was doing, and with Google translate, I could get the gist. I know it's pretty dumb to pick a book by the look of the man on the cover, but you've got to start somewhere. This book had 365 recipes taking in all the seasonal foods. That's how the French do it, now I was too. Every day I looked up what I was supposed to make and went shopping for the ingredients. Luckily, snails were much later in the year.

Our local supermarket had a wide selection of meats. This day I was to try Beef Bourguignon. I picked up some nice looking beef and all the ingredients including the red wine - two bottles 'cause you can never have too much red wine and began my labours. Boomie will eat anything I make. Most of the time I'm a pretty good cook, but this Bourguignon was a little tough, and quite strong in flavour. We couldn't put our finger on what the problem might be, but when Boomie said kangaroo it triggered a small nub of memory in my brain.

I fished out the packaging from the bin and confirmed my suspicions. I'd bought horse, *cheval*.

Some people might eat Trigger (see Roy Rogers and Trigger, his horse), but the Ashwins didn't. The vegetables were delicious, the pasty excellent, but we gave the meat to the dog next door. Pim loved it and so he should, slow cooked in red wine and onions, lovingly stirred and all in a home-made pastry crust. We drank the other bottle of red and vowed to read the labels a little more closely in the future. My cookbook didn't say anything about horse, so it was an experiment we didn't need to repeat.

As the days lengthened and warmed, we decided to go out somewhere once a week, weather permitting. We'd galivanted all over the place, but because the first year in our village was without transport we'd not seen too much of our own Brittany. It was time to redress the balance.

Because we live in the middle of Brittany everything radiates from us. You can go to the Atlantic coast just as easily as the English Channel. You can go to the tip of Brittany near Brest in the West and to Rennes in the East for about the same mileage.

We decided to go south to Lorient on a beautifully sunny day, have lunch and come home.

The President of France, in his wisdom, had changed the speed limit on highways from 90km/h to 80km/h. This wasn't a popular move and the speed cameras all over the country have been neutered. Some are burnt, some are smashed, and the remains are covered in black plastic with hazard tape. The 80km/h limit is just a pain in the butt. On the way to Lorient, there are great straight stretches of wonderfully kept, ultra-smooth roads and we can only do 80km. It's a crying shame when you know 110 might be more appropriate. Most of the roads we had encountered were well maintained, and built to an exacting standard. The French know how to build a road, perhaps they took their mark from the Romans all those years ago. Some sections of the A1 National Highway in Australia are lethal with crumbling verges and potholes. Having said that, Australia does signage well. France isn't far behind.

Lorient was the place the Germans built their U-boat submarine bays. In France, you don't need to go very far to come face to face with modern history. I notched up another walk around France with a fullface helmet as I found a park, then we strolled around in the sunshine, going to the information centre and collecting pamphlets. (I'm incorrigible). One day when I'm old I'll look at those pamphlets and remember our time in France, that's got to be worth something. There is the high road and the low road to get home from Lorient and we took the high road. It winds up and to a radio tower on the top of a hill where the view is for miles and miles. There is also a very, very small bar in the village at the top and we were thirsty.

I don't think anyone, other than the locals visits the bar. Our arrival on a motorbike occasioned some stares, some looks and our every move was clocked.

'English?' the barman asked with a snarl.

'Australienne,' I said and everything changed. The word went around the bar, not hard when it was the size of a 3mt x 3mt gazebo. We were suddenly welcome, best friends, and part of the family. A man with one tooth to his name patted

Boomie on the back and shouted him a drink. I drank it, miming the Gendarmerie and handcuffs, which got a laugh.

'The war,' the barman pointed to our one tooth friend.

'Ah.' We nodded.

When you meet people who have a connection with the war, an interest in your story and are instant friends, it just makes you love France that little bit more. How we wished we had a command of the language to talk to these people.

Our friends in the village have a three-story house and the story goes that in the war, the previous owners were watching the bombing of Lorient from the loft dormer window. That's just over 80km away. That's the impact of war. Those same friends said we should look at a particular balcony at the next village to ours. We took a look at the small plaque. The Germans hanged a young lad of 16 years from the balcony as a warning of the dangers of resistance. When you see things like that, close to home, you hope war never comes again.

I started to research our village and the part it played in the conflict. When we next went walking the landmarks, the forest, the hills took on the stories I had read, the lives they had saved, bringing it down to the level of one man, one woman, one farmhouse, one gun.

We had other friends in the village who were great readers and swapped books if and when we wanted something different. It was these friends who invited us to a BBQ.

Now I'm not saying Australians are the best at a BBQ. The Brazilians do a fair go, and the Americans can give us a run for our money, but Aussies know what's what. Our English friends had a bit of an idea, and the concept was a sound one, it was just the execution that was lacking. It was a small hot plate set in the middle of the table - indoors.

The result was a smoke-filled room and burnt offerings. We threw open all the windows and would not have been surprised to see the *Pompiers* (firefighters) knocking on the door. To add another level of chaos to the lunch, they had a three-legged dog who loved nothing better than to jump on everyone and everything. As we opened the windows, the dog was determined to escape.

'Open the window.'
'Close the window.'
'Open the window.'

To save the day, the dessert was the best bread and butter pudding we have ever encountered, and try as I might to follow her recipe, I can never get it just so.

Our other friends were a mine of information in regard to where we should go, what was worth a visit, what we might like to see. They had been living in the village for years and in their younger days were great hikers.

I took notes and we made plans.

Our next trip would be an overnighter in Vannes. Naturally when you look at the history, the French say in tones of aggrandisement,

'Vannes was settled over 2,000 years ago'. Of course it was.

The day was perfect, the flowers around the outside of the Medieval wall put on a show and the whole things was a picture postcard. Vannes half-timbered houses and narrow cobbled streets inside the walls still hold commerce, houses, restaurants and life continues.

I found a wonderful hat shop. I can fit a beret in my backpack not a problem and visited a print shop that sold unusual postcards, old manuscripts, movie posters and the like. How much does a postcard weigh anyway?

We'd booked a room in an F1 hotel for our overnight stay. These are super cheap rooms, shared bathrooms and all over the country. We always stay in an F1 in Paris, and for the most part, you can always get a room. Our F1 was within walking distance to the town, so we could go out for dinner and drink, then walk home. When you need to ride somewhere, drinking really isn't an option.

We found a small restaurant in a half-timber house from the 1500s and had galettes. These are buckwheat crepes with a savoury filling of your choosing. Sometimes they are called black galettes. They are surprisingly filling. With the local cider to wash it down, we felt we were really French, or Bretons anyway. I managed to order it all in French feeling very proud of myself. We finished with a crepe filled with ice-cream, strawberries and more cream, drizzled with chocolate, life was pretty good. If this was living it up, bring it on. We never wanted to leave France.

And then, we had an email.

Eighteen

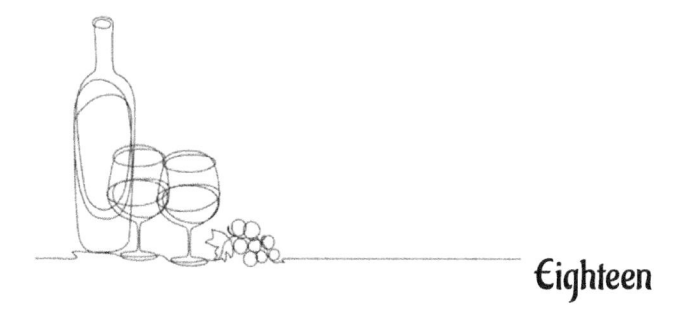

Things intrude when you least expect it. Here we were having a ball, enjoying life and an email changed everything.

Someone wanted to buy our boat.

We didn't think we wanted to sell it, but as we mulled over the offer we began to think seriously about what we wanted to do with the rest of our lives. That is a big think.

Boating had been in our blood for over 15 years, and a few years before that as we renovated the boat before Dikera. We counted ourselves as boaty people and just assumed we would always be boaty sort of people.

But France, our little house, our van, our motorbike all combined to pull at our loyalties. Where would we like to be in 5 years, 1 year?

We sent the fellow an email, saying we were in France, we had a mooring lined up in Mooloolaba, Queensland and what was he putting on the table.

We knew what second-hand boats were worth, practically nothing as the boating scene in Australia was being squeezed dry. When we started the lark there were adventurers all over the place. Over the years the numbers had dwindled to less than half that number. Now it was just too expensive, too hard, too many restrictions, and young ones didn't want to do it. The demographic was mostly over 50.

We looked at his offer.

'You know we will need to go back to Australia.'

'Couldn't one of us go?'

'We should do this together.' The boat was in joint names anyway.

There would be surveyors to organise, a haul out and antifoul to paint. It's not like buying a second-hand car where you hand over the keys and you're done.

We prevaricated for a week, tossing up the pros and cons of the offer, the flight home, the money on the table and the expense if it didn't go to plan.

I emailed our mooring man and asked if it was possible to bring the date forward just in case the sale fell through. Then we'd be able to sail south and stick to the original plan.

It turned out that the boat on the Mooloolaba dock was leaving early.

That was one domino out of the way. I contacted the slipway and asked about the availability of a haul out. We had a few dates to choose from that coincided with the tides. Another domino fell.

We got in touch with a surveyor and he was free on our haul-out dates. It was all coming together too easily. If you believe in fate, it was meant to be.

The last thing I did was ask for a deposit from our man. We didn't want a tyre kicker. It's not like we could just pop around the corner and be back for tea. We would be travelling half-way around the world and back.

With everything in place, I purchased airline tickets.

But just maybe we could be back for the rest of the summer ... after we'd got rid of all our boat stuff, our belongings, our life for the last 13 years.

I felt like a seasoned traveller as we made our way, yet again, to Charles de Gaulle airport. I'd given us one day in Paris and we used it to stroll the streets, immerse ourselves in all things French and I visited the Place Vendome. I'd just read a book on the history of the Ritz Hotel on Place Vendome and wanted to see for myself. I didn't see anyone famous coming out but there were a number of well-heeled people going inside. How the other half live is always good for a bit of a gawk.

Paris has more than one face to show the visitor. There are all the usual places, the Tuilleries, the Champs Elysées and the river. I'd read Les Miserable too and this fuelled my interest for the street names and places. But the other side is the squatters quarters, the beggars in the metro, the squalor on the railway lines and the no-go zones around St Denis. All big cities have their problems, and Paris is no different, but it felt like we had to take that bit more care of our safety, that extra step to be vigilant.

The tropics in winter are pleasant. For around 9 months of the year the weather is 26 to 28 degrees, not much rain and sunny. Summer is another form of hell.

We arrived back and found a lift to our boat. She was as we left her, but a bit mouldy on the inside. By my reckoning, we had two days and then the tide would be right for our haul out. We could shift all our stuff off the boat while on the hardstand if the prospective owner liked what he saw.

He liked what he saw.

It just about broke my heart when he said yes. All he was waiting for was the surveyor's report. He was very excited about the prospect of sailing our boat and I knew that feeling because I'd had it when we bought the boat all those years ago. There is a great expectation of excitement and adventure. A dynamic melding of man and the sea. I could see in his eyes he was smitten, just like we'd been smitten. It wouldn't matter if the surveyor had said she's a leaky old tub and I wouldn't touch her with a barge pole, our man would still want her. He was in love.

I felt like I was giving away my first-born.

We hired a lock-up and pulled our stuff off the boat, sorting our life into what we needed, what we wanted and what the new owner could keep. It turned out we didn't need anything at all, and in the end, we just kept our personal items and a few mementos. Just enough to fit in a suitcase. Not much to show for years on the sea.

Now we didn't have anything in Australia except a bank account. Not a thing.

But, we did have a healthy bank account. There is always an up-side.

Nineteen

With the money from the sale of the boat, it would be time to do some more renovations to smarten the old place up a bit. We could finish the bathroom, install the new toilet, strip wallpaper and paint. If we worked at it for the rest of the year it might be finished in time for the biking season next year. That was the plan.

We'd missed out some of the summer months while selling the boat, but autumn is the season that brings crisp fine days, blue skies and the colour of the turning leaves is magnificent. I missed the seasons in the tropics. Autumn also meant I could indulge in knitting. Hats, jumpers and beanies were the order of the day. Boomie and I had stocked up on cheap wool in Australia knowing I'd be knitting over winter. When you have practically lived in shorts and t-shirts and 28 to 30 degrees for years a cardigan or jumper (sweater) is the last thing you have in your wardrobe.

With something to look forward to, we got stuck into the house.

The French have a love affair with wallpaper. All our rooms had wallpaper of varying degrees of hideousness. Some had more than one layer.

I'm one of those people that like to pick, winkle out and strip wallpaper, although I didn't know the latter until I started, having never stripped wallpaper before. We'd purchased a stripper in Australia and it had been sitting with our tools just waiting to show its muscles.

To do the wall properly you need to get rid of the skirting boards. They were a mismatch of wood, styles and such, so Boomie brought out Crowbar man. He is the sort of person that likes to name inanimate objects. I think it's a man thing.

So Crowbar man got to work. There was quite a bit of renting and splitting. Upside? We could burn the skirtings in our heater. Lead paint?

'Um, I dunno.'

And didn't we find things that had, over the years, skitted under the skirtings. French Francs from the 1930s, millions of dressmaking pins, hair clips, razor blades, pencil stubs, a mouse nest with newspaper from 1967 and fluff balls that would give a fur ball a run for its money.

One coin in particular was 10 centimes with a hole in the middle. It was Art Nouveau in style and I wore it around my neck on a piece of string. Myriam's son collects coins and we had given him some Australian money which went down a treat at his school's show and tell. I asked him if my find was worth anything. There is a web site where you can find out these things.

Giullaume found I was, quite conceivably wearing 125€ around my neck. That was the price for perfect, no blemishes. Not bad for a mornings work!

Now I could get stuck into stripping. I started in our bedroom. Behind the wallpaper I found pencil notations on the paintwork. Measurements from a carpenter and a note to say they would be back on Monday to finish. The previous owners, Louis and Bernadette Cansot would have read that scribbled note on the wall. We'd found Louis and Bernadette's graves in the local cemetery, along with their first daughter. Louis we were told was the local postman until he died and Bernadette had about 30 years as a widow. We don't know how she died, but we could see she was 93 which gave a good indication she was just worn out. I did find a few clues around the house that got my imagination working overtime. The front door had been smashed in with a boot at some point and a new lock fitted and the splintered door frame repaired. And there was a dark splotch on the floorboards near where she had had her bed.

And the evidence suggests M'lud, that the woman got out of bed, slipped, hit her head on the end of the bed, bled profusely and when she was found missing, the door smashed in for a rescue.

Bernadette had a reputation around the village as feisty, not to be trifled with and a bit of a grump. She ended her days in the Government-run old people's home, conveniently situated opposite the cemetery. I liked my scenario, it fitted the clues. Boomie said the stain looked like some sort of solvent and the wooden boards had soaked it up. My idea was much more inventive, I'm a writer after all!

Wallpaper stripping is satisfying once you get the knack of bringing down big sheets of paper. I can understand why demolishing a building has such an attraction.

After I'd done our bedroom and the walls were cleaned of the old glue, Boomie and I got to work on the lambris. This is like panelling that goes half-way up the wall from the floor and is capped with a fancy bit of timber. It hides a multitude of sins. And we had lots of sins. The plaster in the house was very chalky. Probably just after the war they didn't have all the good stuff and made do with whatever they could lay their hands on. Our plaster was mainly chalk, talc and a prayer. You only needed to touch it with a scraper and it would bruise. You could dig a hole in a second with a toothpick. So, to cover the lower half of the wall was an excellent idea. We got to work, using our boat building skills to cut sections of the lambris to length and stick them on. Lambris is a popular choice in France as most people have the same problem as us. The hardware store have cheap, MDF covered in printed (fake wood pattern) paper, they have the MDF with a veneer of wood and then they have wood. It's tongue and groove so slots together. We went for the cheap variety, and as long as it didn't get wet it would look good and last many years. With our new heater, it wasn't likely to get wet.

I think autumn is the time of year people renovate. It was a race to get packs of lambris as it flew off the shelves. We nearly had an altercation as a fellow wanted what we wanted, but we acquiesced with gracious good humour … well, Boomie held me back, the voice of reason. It was already on our trolley for heaven's sake and we'd left our trolley to grab some glue and the fellow should have seen it was on our trolley.

After getting all our lengths cut we decided to paint the walls, then stick up the panelling; it would save a lot of masking up.

The French, we found have a unique way with paint. In Australia, you pick the base paint off the shelf and take it to the desk to be tinted with your chosen colour via a colour chart. It's computer-generated and you can be sure of the same colour, pretty much most of the time.

Our local Brico (hardware store) have pre-tinted paint in 1lt and 2lt and then you can buy the base and the tint and do it yourself. A few drops here, a few drops there. There are tinting stations around the place but they didn't have the jiggle machine to mix it.

We decided to try the DIY method.

I'd picked a dye that was sort of dark milk tea. It came out looking like baby sick. We added some dye from another tube we'd bought for the bathroom, a grey which was in reality a blue. Our baby sick went purple.

'Ah, what the heck,' Boomie squirted in all the rest of the milk tea and we ended up with a nice pastel maroon that reminded me of the Victorian era and old-world roses. It suited the room perfectly and miracles of miracles it was a perfect match for the curtain material I'd bought in Australia about 2 years ago. I'm talking a perfect match. We couldn't have done it better. We could never repeat it.

Our ceilings were of the same poor quality plaster so we decided to use the 1970s look and go for stipple. It hides everything. The paint we bought weighed about 15 kilos and to add another 15 kilos to the ceiling was a bit of a worry, but we pressed ahead. The paint was mainly sand and flicked everywhere, but covered all the small imperfections and cracks. It dried to a cream that matched our wall paint which matched our curtains and toned well with our lambris. This decorating lark was easy.

I was the paint mixer and hold everything, Boomie was on the ladder and we got to work. In one afternoon we'd done the room and it did look special. Such a difference from the faded ghastly wallpaper of pink, green and grey flowers.

Our lambris came next and in two days Boomie had completed the walls. We bought new skirtings and with a new light fitting - one room down 4 to go.

I'd seen a sale for paint in our junk mail and we went shopping. We opted for a pastel grey for our front room, and the paint tub indicated pastel grey, but when we opened it, baby blue stared at us. So the room would be baby blue. With another 15kilos of paint on the ceiling and cream trims, with silver-grey curtains that room was complete, I just needed to get into the roof space to wire up a new light fitting.

The beams in the roof were as thin as the beams under the house. This necessitated crawl board to be positioned where I needed to go. Up in the roof we only had 4 boards so it was one over the other to get across to the right area. One wrong move and your foot would go through the ceiling. I'd been up there before and didn't like it much, but if we wanted our new light … I began the perilous journey across the ceiling to the far side of the house. Just to make matters worse there is itchy insulation over the beams so you are never sure where they are. It's a long and torturous trip.

'Are you there yet,' Boomie would call.

'Getting there.' I had to lie down on my board to get to the spot.

I pulled back the insulation mat and thousands, or it seemed to me at the time, thousands of little spiders began to crawl over my hand and arm. The mother spider was none too happy either. They were jet black, about the size of the end of a pencil and it took all my self-control not to scream. Little babies everywhere. If it was Australia I would have screamed, but as far as I knew there are no poisonous spiders in France. I waited until they moved on, then got to work, drilling a hole for the anchor hook, wiring up the light and then, backing out back to solid ground.

As soon as I landed on the grass, I threw off my clothes looking for spiders, much to the delight of Giullaume next door, he's only 10 and finds everything funny.

In-between renovations we went out and about visiting the Nantes-Brest canal for long walks along the towpath and hunting out markets. I love markets. You can sometimes pick up unusual objects, rummage around for old jewellery and books. Boomie likes them for the tools and machinery that are on offer.

There is a big market once a month near us that Myriam goes to with her stall of 'junk'. She sells bits of furniture, wicker baskets, old books, toys, and anything she can find that people give her. She had, which we bought en masse, original pages from old c1920 illustrated French magazines. There were adverts for Chanel perfume, cars, radiators and fans and tourist holiday destinations. All original pages in the most magnificent art deco drawings. We had the idea to frame them for the house.

With Myriam's directions, we headed off and found the, way to the village of Maël-Pestivien.

This market has everything. A lot of English people have stalls as well as Gypsy types and the ordinary French. You can buy a circa WWI rifle as easily as a set of Lego or an 1833 postcard of Paris. It's wonderful.

Boomie bought a pig-skin coat for 10€. Super warm. Later we had it cleaned for 80€, so not the bargain we imagined, but you do these things only once, and it all add to the flavour of the story.

I had my eye on a 1928 Art Deco clock bookended with leopards all in green marble. I did buy an English cookbook 1934 with weird and wonderful recipes. Invalid broth, fish head glue and how to iron a shirt. Books like that are fun to read and who knows, maybe one day I will need invalid broth.

We also picked up an ink pen/letter opener made of bone. It had a tiny eye-piece in the body and if you look inside it shows you the main figures of the First World War.

George V, Nicolas II and Albert I in a small pen and ink drawing. I Googled the object and found they were small souvenirs sold during the war.

The market also does a good sausage in a baguette with beer. Just the ticket on a frosty autumn morning.

My bucket list included the standing stones at Carnac, so we decided that would be our next overnighter.

The whole area around Vannes and the *Baie de Quiberon* is littered with Neolithic sites. And who wouldn't want to live there with the sea, and beautiful countryside? These people are the later part of the stone age 12,000 years ago when farming and animal husbandry were the new thing on the block.

The Carnac stones are, so the museum told us, the largest collection of megalithic standing stones in the world. Of course they are!

3,000 prehistoric stones were cut from local rocks. I was dying to see them for myself, having studied archaeology and worked in the field.

Not only were we the only people in the early morning to be at the sites, but the fog was just lifting and the sun was peaking through. Perfect atmosphere for some photographs.

Naturally, my mind was working overtime thinking on these ancient people co-operating to put up these lines of megaliths. They had something in mind, but what it was, is lost to us, although some boffins suggest with certainty that the *Alignement de Carnac* are to mark territorial ownership. I preferred my conjecture that they were to mark family groupings, but don't ask me why. It's probably my overactive imagination. Did I mention I was a writer.

Whatever they had in mind, the lines, the dolmens, the tumuli are all great feats of engineering in an era when they were still collecting and napping flint.

The visitors centre is quite good too and being the only ones there, we could wander at our leisure. They have a great collection of old photographs about the first French people who studied the stones, who dug around them, conducted archaeological digs and surveyed them. You can never say discovered, because they have been part of the landscape of Brittany for a long time, but to have the time to get to understand them only comes with free time, which a peasant in the middle ages probably didn't have.

Some of the stones have been incorporated into dwellings and houses which is interesting too and I saw one that was the outside of a chimney stack on a house. Once again, when you live with something day in, day out it ceases to become a wonder and soon becomes the ordinary.

'Oh, this old thing, just a megalith from the stone age, pfft.'

A dolmen is a funeral construction formed of a horizontal slab held up by two upright slabs. The oldest European examples are in Brittany, and date to the 5th millennium BCE. Naturally!

Sometimes they are covered with dirt to create a mound. We hunted one such mound off the beaten track. It was set back in a wood and you needed to stoop to enter the alley, then in the centre was an enormous single slab for a roof. It was cold, dark and makes you think of the people who had the idea in the first place.

'I know, let's build a big thing.'

'Yeah, sounds like a great idea'.

Breton dolmens date back to 4500 BCE, earlier than the Egyptian Pyramids which puts the effort into perspective, doesn't it?

Our decision was to stay at Auray and we snagged a room in a pub, right on the quayside. It fitted the adjective of charming. We had a huge fluffy quilt, little lead-light windows, a winding oak stairway to get to the back door which was about 4ft tall. There was a pub next door which served meals. Perfect.

Auray is at the head of the Loch River and has several boasts, besides its charm.

The ports strategic position meant it collected duties from boats passing through which made it wealthy. Never miss an opportunity to tax the working man. In the 16th and 17th century grain and wine, so the pamphlet said, made it the third most important port in Brittany, and just to add a little more aggrandizement, Benjamin Franklin landed there in 1776 to meet King Louis XVI.

We found it lovely.

Twenty

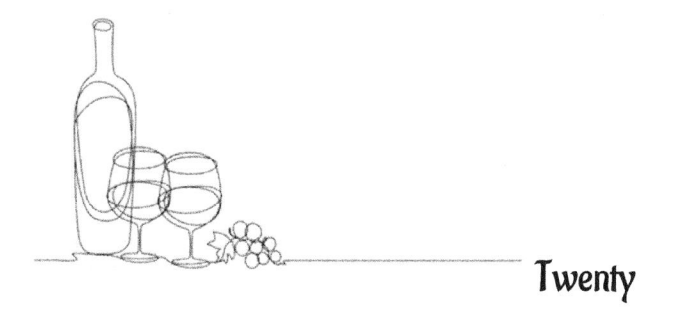

As the season wound down and winter was fast approaching wood again was the order of the day.

The Brico in St Brieuc had wood at a reasonable price, but the delivery to our village was very expensive. I calculated what we could fit in the little van.

Buying 1.3 cubic metres of wood doesn't sound a lot. Seeing it in the car park, it doesn't look that much. Putting it in the van is another matter.

The man who used the forklift to bring it to our van was sceptical. He scratched his head and other body parts and gave a gallic shrug.

The trick was to go high and build it towards the door. The van began to get lower and lower on its axels. Some people were amused at the Ashwins, some were wide-eyed at our stupidity. I was going to make that wood fit if it killed me. It nearly did as my back was bent double lugging 30cm logs.

But, it fitted. With a small bit of jiggling, we could close the doors. We just saved ourselves 50€, which prompted us to spend the money on wine. Win, win situation.

Now to get it out again and stacked on our woodpile under the house. As I said before, when you have wood, you become a glutton. We set the fire and fed it constantly and really it doesn't feel like burning money.

They say owning a boat is like standing under a cold shower ripping up $100 notes. Owning a wood stove they are only 20€ notes. The Ashwins can put a positive spin on anything, and we did spin it, every night as we sat in front of our stove, warming our toes, drinking our wine and congratulating ourselves on living it up in France.

That sort of talk can be addictive.

Of course, nothing comes without a cost. Boomie received a speeding ticket in the post. Apparently, we were pinged doing 80km/h in a 70km/h zone near Rouen. The fine was 45€ and one point off his licence. Well, considering we'd been waiting for our swap of Australian licence for French licence for over 18 months we were not quite sure where the point would be deducted. But, on the other hand, the letter said if he was a good boy for 6 months they would reinstate the point. Again, where we didn't know. I didn't think we'd ever get our French licences, because when we put them into the Prefecture it was the exact time they were going on-line for all licence paperwork. Our application was probably in the bin. We still had our Australian licences, and Boomie had his international that we'd recently renewed, but there would come a day … when we'd need to reapply.

Our shopping expeditions are never far from a hardware store, no matter how hard I try to stay away. It was in the Christmas sales at the Brico I found a flat pack kitchen with 30% off. I'd been carrying the dimensions of the kitchen around on a scrap of paper for months in case we saw something and here it was.

We bought it, then purchased an oven and a stovetop to go with it, a new sink and taps and had something to do over the Christmas break when everything is shut.

It's a bit of a rip off when you buy a kitchen and then need to throw half the benchtop away because of the cut outs for the sink and stove. This one gave us, just enough space for the screws, and just enough is good enough.

We put it away until Christmas Day.

I don't know about anybody else, but if one thing is to set the road to divorce, it is assembling a flat pack. We'd done this sort of thing before, and knew all the pitfalls, all the traps, all the grinding of teeth and pursing of lips. It's bloody hard not to say something when you can see - that needs to go there, blind Freddy can see - that needs to go there, but the love of your life can't see it. On the other hand, he can tell you 'till he's blue in the face that the screw won't fit and the People's republic of China didn't include the right fitting, but would I believe it?

We made a pact.

Slow, easy and take a break.

Just as I can read a map and find directions a doddle, I can read instructions. We got to work and in no time, we had a carcass, then doors, then drawers.

Getting it in the kitchen past the tight curve of the front door and atrium was a worry, but with some heaving and taking the front door off, we had it inside.

In preparation, we'd demolished the old kitchen bench and I'd used a can of degreaser to get the walls clean. Everything was ready for the first fitting. It didn't fit.

'Merde.'

The problem was the floor. If we put the cupboard into the corner it wouldn't straddle the floorboard dip and wasn't level. The solution was to leave a sizable gap on the wall end so that the two ends of the carcass were the same height and there would be a 10cm gap under the middle. It wasn't ideal but it was the only thing we could do with the huge bow in the floor, courtesy of the sagging beams underneath. If the thing didn't sit square, none of the doors would shut and the splashback would look wonky.

After getting it right, the oven went in without a hitch, the sink was another matter. When we opened the box, half the fittings were missing and there was a dent in the side. We took it back, as with all big hardware stores, they didn't blink as the man was sent to get another. When we opened this one, it was obvious, our previous purchase had been installed and uninstalled. There was none of the protective plastic, the little bag of bits and booklet.

The plumbing was a trial, because Louis Cansot, the handyman extraordinaire had done a very creative installation which when taken apart left us with fittings that didn't fit any modern pipe fittings at all. Our bargain kitchen was now costing a bit more than 30% off. We acquainted ourselves with French plumbing.

Now, it's a bit of an admission, but I don't like plumbing. I quite like doing a bit of electrics, painting, sanding, stripping wallpaper, assembling kitchens, but plumbing leaves me cold. On the boat, there is quite a bit of plumbing as things need to be routed through bulkheads, connected to pumps and valves, and I guess I've just had my fair share of one-way valves and hose clamps.

'Shall we call an expert?' I asked.

Boomie looked at the problem. He's quite good at plumbing.

'I think I can do this.'

It turned out Louis's inventiveness had wrecked the small bit of pipe we needed to be able to join A to B to flow to C etc. We called in a plumber.

They came with brazing tools, all the U-bends and olives a girl could want and got the job done in under an hour.

'You know, if we had the money, I'd get a little man to do all the renovations.'

'Me too.'

But we didn't, and we couldn't, and so we had to do it ourselves.

It is just fabulous to work in a new kitchen. Everything is so clean and well ... new. To celebrate, I wanted to make something nice.

Our previous oven on our cheap stove was christened the crematorium. It didn't have a thermostat that was worth two cents and I needed to watch my food constantly. This new oven had all the bells and whistles - for the price. So I decided to make a meat pie. We like a pie now and again. It's an Aussie thing.

I was braising the meat and somehow it just didn't smell right. Not off, but strong, but not strong like horse, just iron sort of blood strong. I have a nose for food that's iffy.

And when Boomie came in and said something stinks, I knew I was right. With my suspicions aroused I fished the packaging out of the bin.

Aliments pour Animaux.

I was cooking meat for animals. The thing about the mix up was that the meat sat right in the cold department with the human consumption meat. It looked just like human meat, it looked really good quality meat.

Pim, the dog next door, had a good dinner that night. Giullaume thought it hilarious.

Ya win some, ya lose some. Don't ask me what sort of beast the meat came from, I have no idea, and I'd rather be left in the dark.

We had pizza instead.

I'd bought a chicken once that said corn fed and it was in the sell-by fridge which is always a bargain. It was the toughest thing I've ever tried to cook. I found out it was a rooster. I have no notion on what the French do with a rooster other than boil the thing for about 6 hours. Never again.

Our supermarket carries a lot of things that are for the demographic of the neighbourhood. Old people like the staples they grew up with, so we see all manner of offal, fish bits, blood sausages, and half a calf's head. They do a roaring trade in tripe, brains, kidneys, and fat. You can buy a great lump of lardy fat that on the packet recommends roasting. It's not crackling, but just fat. My cookbook thankfully doesn't mention fat. You can also buy duck fat for the discerning cook. Heart attack material in a jar.

But what the French do around where we live and I suspect all over, is eat seasonally.

And in Brittany, the women go in for preserving in a big way. As the season for fruit approaches the shops begin to sell jars with sealing lids, big pot-boilers, thermometers and all the paraphernalia for preserving. It makes you want to have a go yourself, but I held back. It's a lot of work. My family did it when I was a girl and it's not fun.

But the local women swoop on boxes of strawberries and stone fruit then get to work. We gave Myriam's mother all Bernadette's jars when we moved in and were cleaning up.

Eating seasonally makes you shop differently. We took a liking to endives in bacon and cream sauce. I bought endives every other day while we could. The same goes for mushrooms, all varieties. Then it was the turn of fresh shellfish. Then apples.

Christmas time brings the exotic from Africa and beyond. They had mangoes, limes, pineapples and persimmons.

I like walnut season. They taste different when picked locally. I could crack a nut and eat it with one hand, and we burnt all the shells so they did double duty.

Over the winter months we didn't use the bike, my long underwear wasn't for sub-arctic conditions, so we drove around to visit places, have hot mulled wine, eat pom frites at the beach and walk tracks in the forest.

We did have one or two overnight stops and Brest was a highlight. Brittany juts out into the Atlantic and Brest is right at the west end of France. It is on the river Penfeld and has been a safe port and haven for generations of fishermen and sailors. As you walk around you can see the town was built around the water, with buildings and walls to weather the weather. It's been strategic for years and in the Maritime Museum we found that Cardinal de Richelieu decided to make Brest a major naval base in 1631, the Germans had a go at occupation in 1940 and built submarine pens and WWII just about destroyed it. It was also the first port of call for the Americans as they sent their boys to the front in WWI. In winter we had all the tourist attractions to ourselves, which is the best way to have them. You can get good photographs, you don't get jostled or moved along. I'd booked a hotel over the river inlet and it was when we were coming back after a day of exploration that things got a bit mixed up.

There is a movement called the Gilets Jaunes (yellow vests) which started in October 2018 protesting the

Governments increased fuel prices and high cost of living. It gained ground and they were in full swing when we were in Brest. The protestors blocked the only bridge over the inlet forcing the motorists to detour. Most of the public were good-humoured about the interruption and toots of support were very common, but the detour was torturous. The next bridge was 60km away and as the cars were banking up, I wondered if we get to sleep in our hotel room at all. There was a huge traffic jam to get over the tiny bridge in an even tinier village. I've never been so glad of our GPS. It was dark on the back roads once we'd passed the bridge crossing and we needed to head back into Brest to get to our room, except we were off the highway and no way of getting on. You could go under it, over it, beside it, but not on it unless you went over the blocked bridge again. It was only with some inventive driving when we went off-road, we found our way back, about 4 hours later. We'd missed our dinner, all the shops were shut and so opted for a vending machine special. Chocolate is food!

Saturday was the designated day for protests, so going out and about mid-week was the smart play.

There's an Abbey near us that we kept promising to visit and it was a cold, blustery day when we decided to finally see it. It's often the way that you will travel miles for something, but if it's on your doorstep you never get around to it.

Bon Repos Abbey was first built in 1203 and the pamphlet says it one of the largest abbey churches in Brittany ... naturally.

It has suffered over the years from neglect, lack of money and took a beating in the revolution, so much so monastic life finished there and the property was sold. The next time it came into use was in Napoleon's time and it was used as accommodation by the engineers working on the Nantes-

Brest canal. Now it is privately owned and probably a huge money pit. The entrance fee wasn't expensive and the upkeep would be enormous. I guess it would be up there with owning a boat, a horse or buying firewood.

We walked around the grounds trying to envisage how cold it would be as a 13th Century monk with sandals. That's real dedication to your faith. The Abbey was responsible for its own food, then its charity to the surrounding people. They say it gained great wealth and civic responsibilities, which translates into power over the minions. The threat of hell's fire and damnation for not paying taxes was always a good bet. Most of it is in disrepair, although the volunteers valiantly try in the summer months to keep it going, it looked like they were losing the battle. Once you get weeds in the blockwork it doesn't take long for the mortar to crumble.

I'd seen another Abbey on the way to the Granite Rose coast so decided that would be our next trip.

Abbey de Beauport is part ruin, part museum and the two meld together very well. This was once a wealthy hub of activity constructed in the 13th century. I liked it for its defined layout, stunning view of the estuary and the bay of Paimpol. It also had a, still producing, orchard of apple trees. These trees were ancient varieties and the monks made cider from the fruit. The sun was shining, albeit cold, we could see the everyday life of the monks, the kitchens, the back steps and latrines. That's the sort of detail that brings things to life.

It too suffered in the French Revolution and was closed and sold off to three families. Such is life.

Back at work on the house, we looked for any excuse to knock off and have fun. It came with an invitation to our friend's house for lunch.

'Not another BBQ I hope,' Boomie said.

This time it was a table-top raclette grill. The French have a love affair with cheese and especially melted cheese.

We were to be treated to a typically French affair where you melt your little bit of cheese, or very big bit as it turned out, on a small shovel over a flame thing then pour it on your chosen bread or vegetable. Sort of fat on bread, melted fat on bread.

I love cheese, but even this had me beat. The raclette cheese is super gooey and super loaded with fat. It was originally Swiss, but the French have made it their own.

'I'll probably need a by-pass after this,' I said to Boomie.

The French suggest white wine to go with the meal as they think that water will harden the cheese in your stomach and cause indigestion - if you live that long.

The trifle for dessert was a masterpiece. I would have skipped the cheese and gone straight for dessert. Our cook really knew how to make desserts.

I promised a reciprocal dinner in the future, thinking a quiche and salad. These people were English, and as the lunch progressed talk came around to Brexit. They were living in the European Union without visas, without any restrictions or conditions as European citizens. As Australians we were loaded with dos and don'ts. What would happen with Brexit and the English who had made France their home was on everyone's mind.

Because we'd had a bit of a palaver the year before regarding our visas I was determined to get things moving early. A wise choice in hindsight.

Twenty one

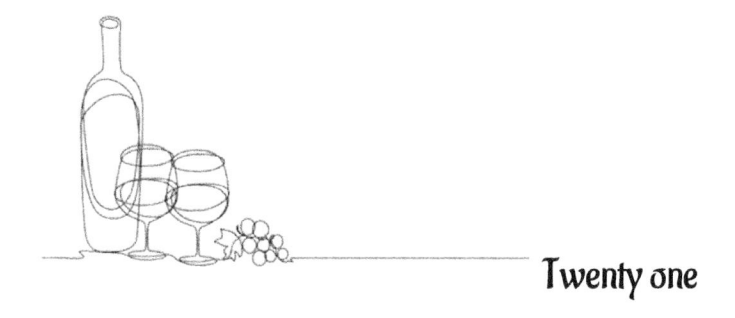

You can apply for a renewal three months before your visa expires. That was the guideline we had used previously. So, I emailed our intention to get an appointment and all hell broke loose. The Prefecture site was down.

Not a problem. I gave it 24 hours and tried again. Nothing. I tried a third time and the whole system of appointments had changed. What I thought I knew was as much use as a chocolate teapot.

Now I had to get a slot for an appointment. Previously we'd always gone in for our interview together. Now I would need two appointments. All the spaces were greyed out. There wasn't one free spot to be had. I wasn't going to panic, after all, I had three months up my sleeve. There must be some sort of mistake, a new system, computer glitch, I could wait.

I emailed the office and received a curt response to say I could only apply online at a maximum of 2 months before my visa expired. I would need to wait.

On the very day I became eligible I went online and it was booked solid. I did start to panic, just a little. I asked Myriam to ring the Prefecture for me and the woman on the other end hung up on Myriam. I tried several times to get an appointment with no luck. The days ticked off and we couldn't get near an available slot.

'We need to go in,' Boomie said.

I gathered all our documents in the expectation we might get an appointment, but I didn't hold out much hope, because the Prefecture can be sticklers for procedures.

And it was then, all became apparent.

The English were clogging the system to the extent that it was almost broken. In the rush to get visas like ours as a safety net in case Brexit left them high and dry they were flooding into offices across the country applying as fast as they could type. The system was tight at the best of times, but this was unprecedented and it seemed unforeseen.

The woman behind the glass was at her wit's end. Frazzled and curt she gabbled something of which I only heard two words, Brexit and on-line. I can, sort of get what people are saying if it is clear and slow. I asked again and she waved me away with her hand.

'Just keep trying,' an English woman said as we left.

So I did. Constantly. Every hour or so I would log onto the Prefecture site and give it a go. Someone on Facebook suggested just after midnight. I tried. Someone else on the Facebook group, Aussies in France said they had success just after 9 am. I tried. I kept a diary of my efforts and screen shots as our time was running out.

There is a time-honoured tradition that if you are having trouble with bureaucracy you go to the Maire, the local council. I wrote a letter outlining our efforts and the urgency of our claim and went in search of help.

The woman hadn't heard of the problem, but by the time she had rung the Prefecture and got another woman at her wits end, she knew all the details. By this time the appointment system had collapsed. Chaos.

The French don't do chaos very well. I don't know of any institution that does it well, except maybe communist Poland. I was there in the 70s and if it all gets too hard they just turn their back on you, shut the door and put the kettle on.

The woman at the Maire put the phone down and shook her head. I suggested a letter to say what had transpired that day and get it stamped with something looking official. In dealing with departments, I've found an official-looking stamp works wonders. We had a rubber stamp made of our boat name that looked terribly official and it worked more than once.

With our bit of paper, we came home and tried to figure our next course of action. It was obvious we wouldn't be able to get an appointment. No spots were available no matter what time of the day I tried.

The French have a great love of documents and to facilitate the posting of documents around the country to prove who you are, where you live etc they have an express, registered, receipt envelope system. You get tracking, and you get a receipt back at no cost to the sender to say they have received it. It's foolproof and it is <u>very official.</u> With Myriam's help, I drafted a letter, added my document copies and sent it off. About a week later I had my receipt which I put under lock and key. They surely couldn't ignore us now.

We waited. Nothing. I went back to the Maire and got the woman to ring again to ask what we should do when we expired.

'Wait.'

So we waited. And we became illegal, the only thing keeping us safe was our receipt from our letter.

And while we waited, our wood ran out. The stove had been smoking, like a packet a day man, every time we opened the door. I went to Google and it suggested our chimney might need a clean.

Of course, the smart thing to do would be call a chimney sweep, but how hard could sweeping a chimney be? We'd seen all the stuff you need for sale at the Brico so obviously other people do it. We decided to give it a go.

The first thing we did was mark those pesky chimney pieces in order, so they would go back the same way.

It all comes apart really easily, almost too easily. I took some photos for reference and we gave the chimney a good poke. It wasn't blocked at all and we could see a shimmer of light at the top. With all the pieces on the grass outside, I scrubbed and brushed and put them back together. We reassembled the whole thing, but it was still smoking the room the minute we opened the door. We needed help.

The sweep was a nice fellow. He looked at our problem and pointed,

'What's that?'

'The grate thingy?'

'Yes, that grate thingy.'

'I dunno.'

'That's your problem. No air.'

And he went to work. He had a brilliant vacuum cleaner that was attached to a huge bag that swallowed the stove and it's chimney contents as he put a brush right up the top and twizzled his cocktail stick. I heard tinkling of soot.

'What are you burning?'

'Wood.'

'Nothing else?' he frowned.

'No… nothing.' He didn't need to know about our skirting boards, our furniture, our grocery packaging and anything that had a carbon content.

'Right.'

With the chimney swept he gave us a certificate which we didn't know we needed for our house insurance. Myriam's chimney caught fire the winter before and her insurance didn't pay. I told her to go on tinder, rather than have a house fire to get a date with a fireman.

Instead of buying another load of wood, we made an effort to tough it out with our little kero heater for a month. We could save money for our next trip, and Boomie and I had arctic underwear to keep us warm.

Our next trip away was something we'd wanted to do for many years. It was the sort of thing motorbike enthusiasts say they would do once in their lives.
We were going to the Isle of Man TT motorbike race in the middle of May to the beginning of June. I booked the ferry to Ireland, then the ferry from Belfast to Douglas on the IOM and the ferry off the Isle to Liverpool. What an adventure. What a trip.
But, before our trip to the Isle of Man, we decided to go to Verdun - his bucket list, via the champagne district - my bucket list. We'd have 10 days at home between adventures.
We could go later to Verdun, but all the books we'd read and all the information we gathered said summer is high season, high crowds and high prices. Going just at the end of April, beginning of May we'd miss most of that mayhem.
This would be another bike ride right across France ending in Belgium at Ypres. It sounds a long way but from

our village to the border is about 680km. When we lived in Far North Queensland we'd travel 500 to see family for the weekend, although there is a vast difference between the roads in FNQ with nothing to see for 300km and a road in France that might travel through a town, village or city every 35km or so. I'd calculated we would do nearly 2,000km flitting from here to there and back again. If our appointment came in the interim we'd just wing it.

Two days before we were due to depart we received an email from the Prefecture. We had two appointments. We had one day's notice.

I gathered all our documents together for our 12:30 appointment. The only problem I could see was the gate closed at 11:30 and there wasn't a reception you could call, a buzzer or any way to tell the people on the inside you were on the outside.

It turned out we were a block appointment with about a dozen other hopefuls. We all milled about until a woman came out and led us through the back way to the reception room, past the staff room and the photocopier. We went through the procedure of fingerprints, documents scanned, questionnaires filled in and photographs embossed with an official-looking seal. This time I could carry on a small conversation with her and she explained they were flat out with Brexit people. It was *impossible.* Tell me about it!

But, we were legal. We were set for another year and this time we were told we'd need to take 100 hours of registered French language lessons. There would be a test.

I knew how to swear, how to talk to a plumber and how to read a recipe. Boomie knew how to read a wine label. It probably wasn't going to be enough.

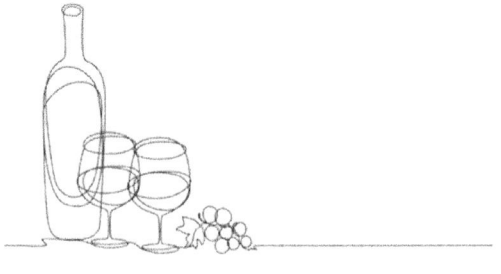

Twenty two

Once again I booked our hotels in advance. I didn't want a repeat of last time when we were caught short. Boomie worked out our route, and I worked out where to stay taking into account our budget, and how much time we'd need to see things - and in France, there is a lot to see.

The first day we'd be heading 400km to Chartres and the weather forecast was fine all the way. This time I left room for pamphlets. Who needs a change of socks anyway.

There is a theory that if you ride often you become 'ride fit'. Sitting in the saddle for more than an hour is about all the ride fitness the Ashwins can endure when we need to get off, stretch, walk around and have a toilet stop. It is supposed to be fun, not an endurance test. There are a lot of roadhouse stops on the motorways and we like them all. It may take a little longer to get where we want to go, but who cares. Our hotel is booked, we are not in a hurry and it makes the whole travel thing more enjoyable.

I'd written out my cheat sheet and we found our hotel in Chartres without one U-turn, which was a first.

We rode straight into town on our first day and I walked around looking for a parking spot. (I'll get that gold medal if it kills me). I saw a woman parking asked her if the park I'd chosen was free because I couldn't see any signs and she answered in perfect English. She was a University Professor who taught English and we soon got talking.

No, the park wasn't free, but she wanted to know all about us.

'Are you Australian?'

'Oui.'

Boomie rode up and we chatted about our travels, our experiences and she said if ever we wanted to come to Chartres again here was her telephone number. What a nice woman. Her family had come from the West Indies many years ago and made France their home. Benita, and people like her make memories worth keeping.

Chartres is only 90km from Paris and has been fought over by just about everyone from the Romans, the English and the Germans. The cathedral which was started in 1193 and completed in 1250 is on a hill overlooking the old town and as you walk up to it the sight is impressive. After seeing so many Gothic cathedrals, you might think we'd be,

'Uh, another church.'

I never get sick of seeing architecture on a grand scale, built in the 12th century when it was a bit harder to achieve results. Chartres cathedral has all the majesty of the others we'd seen, but it was situated within walled gardens and ramparts within the old town and crowded by houses and buildings, just as it might have been back in the day.

It has a relic that people from all over the world come to see. It is supposed to be the tunic, the *Sancta Camisa,* worn by Mary at the birth of Christ. I say supposed because I've seen Saints fingers, toes, relics, the splinter of wood, the facial hair of Mohammed and all the rest and it is a long stretch of the imagination to think these things survived. I'd read stories of monks fabricating Saints digits so they could charge pilgrims, and of miracles that 'happened' once a week to fool the brethren. I'd seen two tombs in Jerusalem purported to be where Jesus rolled back the stone. John the Baptist had about 15 fingers. Nevertheless, I leave it up to the individual.

We walked around inside and out then strolled the old streets peeking into alleyways and climbing steps that were worn down by hundreds of years and hundreds of feet.

I could have snapped of hundreds of photographs in Chartres as we wandered around the canals and the old town. Everything was so well kept. I particularly liked the old wash stations on the water. The pamphlet I'd collected said that they were from the 1500s. We found a coffee shop, stopped and just enjoyed the view. The apple pastry we'd bought just added to the moment.

Our next stop going East was Château Fontainebleau. Just the name makes you sit up and take note.

It was only a ride of 114km from Chartres and from there we were going on to Troyes another 115km. This was the life.

The Royal Château de Fontainebleau is where the Kings of France had fun. It was the only residence lived in by every French monarch for nigh on eight centuries. Thirty-four Kings and two Emperors. With 1,500 rooms, I read in the pamphlet, it is the biggest château in France. Of course it is.

We arrived on a Tuesday. It's shut on a Tuesday.

'Merde.'

The manicured and regal grounds were open to visitors and they are quite spectacular. What was of interest to me was the graffiti scratched into the walls. I saw dates of

1959,1926,1898,1863 and 1702. People never change. What was happening in 1702 that L.S had the time to scratch his initials in a palace wall and get away with it.

After a good wander, we went for a late breakfast in a café across the road. I ordered what I thought was a cheese toastie and hot chocolate. It was melted cheese, both sides, inside and outside. I was defeated by fat, yet again, albeit tasty.

On to Troyes which I knew nothing about and was a great surprise. It's medieval, half-timber houses in the city centre are amazingly well kept and everywhere. It's like stepping back in time. They are all used today for shops, cafes, bars, tabacs and restaurants. The continuity of occupation is fabulous. It is at the tip of the champagne district and of course, there are a lot of premises devoted to champagne. They go the extra mile with window displays and you can walk away with a bottle for 1,000€ if that takes your fancy.

Our hotel was right in the centre of town and we could walk everywhere, which meant I didn't need to find a park every time we went out. We found an Irish pub, had our Guinness and then did a tourist trail recommended by the pamphlet I'd picked up at the hotel. I couldn't have imagined a place like Troyes would still exist after all the wars, the revolutions, the invading armies and other catastrophes.

Troyes had several churches, but the one that made the hair stand up on the back of my neck was Eglise Sainte Madeline. It was started in 1120 and doesn't look like it has had many incarnations since. Huge flagstones for the floor and relatively unadorned pillars give the feeling that this church was not just for show. It's not big, and the dimensions felt more down to earth, for the ordinary man. We had the place to ourselves and you can't help but whisper.

We just ambled along in the evening, not too worried if we missed the hosiery museum to look at stockings worn in the fashion of Kings or other attractions. It was relaxing to just meander, everything was just for looking. When it was dark we could see inside some of the ancient houses and marvel at their roof beams, their thatched ceilings and the leadlight windows. I do like peeking to see how the other half live, you know, the ones that might just buy a 1,000€ bottle of champagne.

Our next stop would be Verdun for three and a half days. There was a lot to see.

The name Verdun summons up the First World War, I think, like no other. It is synonymous with 1916, the longest battle in the war and the Western Front. Boomie knew all the details and with him, as my guide, we would see where it all happened.

The town itself is quite small and our hotel was on the outskirts, right next door to a boulangerie, which was handy.

We gathered all our maps and pamphlets for the coming days and made some sort of plan, although the best-laid plans are apt to change.

There are grand monuments to the war, a museum, the ossuary, the trenches and all the French forts that held the line. In the town, we started at the *Citadel souterraine*. This maze of 4km of tunnels was the heart of Verdun's resistance and when the Germans started the offensive it symbolized the resistance of the whole nation. In 1916 the then French President presented medals to some who held the heart of France. Now, they say, Verdun is, with its 26 medals, the most decorated city of France ... of course.

With tickets we joined a small train that takes you through the tunnels and were given ear-pieces to listen to the commentary. The guy said,

'What language?'

'English,' and he handed over our ears. It was, I think in Dutch, and useless to us. Still, we saw all the pictures, and because we knew a bit of the history, we could recognise the main players, the photographs we'd seen in books, and in a way I'm glad I didn't have the commentary, because sometimes they are so French biased you never get the whole picture.

Blinking in the sun after being underground we walked back to town and because it was Sunday the motorbikes were out in force, promenading up and down the waterfront, called London Dock on the river Meuse. Time for a beer and some people watching, or in Boomie's case bike watching. We had to wait for a seat, then pounced on two and ordered beer. The French were out in force, drinking, eating ice-creams and relaxing. It wasn't long before we were talking to some motorcycle people, who were intrigued by our story. They called over some more friends and it was soon a boisterous party. We were invited to houses, to parties, to a wedding, but I rather think it was the beer doing the inviting. How they thought they would be riding and drinking I don't know, but it didn't seem to bother them. We stuck to one beer each and packet of nuts. After we left our new biker friends with kisses, handshakes and pats on the back we walked through the town to the church and the Episcopal palace at the top of the hill. Both were bombed during the war and there are pockmarks, scars and evidence of war all over the buildings. Bullet holes, strafing and chunks missing show how hard-won the battle of Verdun must have been.

Verdun was seen not so much as a strategic stronghold, but the symbolic heart of France and put up a brave fight. On 12th September King George V of England awarded the Military Cross to Verdun one of only two awards of this kind to go to a city or town. The other one was Ypres. We were headed there next.

But first, we were riding to the battlefield just outside Verdun, where trench warfare was a deadly strategy.

To get to all the big monuments from Verdun there is a switchback road up a hill with blind corners. We were riding easy when an idiot came around the bend, overtook a local bus and just missed us by a hairsbreadth. The bus driver slowed and I gave him the thumbs up to let him know we were ok. A near miss reminds you that there are idiots all over the place, you need to be on your guard.

We carried on, in shock, until we found our first stop, Fort Douaumont.

Again the Ashwins were early and had the grounds to ourselves to walk over and explore, Boomie telling me all the details.

After the French were soundly beaten in the Franco-Prussian war of 1871 they decided to fortify their defence and future safety. These forts were strategically placed and held the line.

The Germans in the First World War were acting like tourists once again and so these Forts, it was decided, were upgraded and fortified to once again hold back the invading army.

I read that Fort Douaumont was the highest and largest fort of the 19 which protected the city of Verdun. It would be, wouldn't it!

Earth was heaped on them, turrets for guns mounted and a small village conducted business underground.

We were first in line to go inside when it opened and roamed around without a crowd. There was a hospital, water tanks, everything to make it self-sufficient.

Eventually, the Germans took it, almost single-handed, by a guy names Küntz and the victory was celebrated in Germany because it was huge PR kudos to take the strongest French fortification in existence. Eight months later the French got it back.

What I can't understand is why didn't they just go around these Forts a little to the south out of range of the guns.

The ditch around the outside and the fort is pockmarked from shelling and this sort of landscape is a feature everywhere. Some of the old pictures show not one bush or tree, just craters as far as the eye can see.

All around the area where the battle raged is now called the Red Zone. No building is allowed, and it was planted with forest after the war, never to be disturbed. Bodies still lie in the ground and the scars of bombardments, trenches, and the horrors of war are everywhere, where-ever you look you see overgrown trenches, artillery holes and mounds. The only flat bits of land are the roads winding through the forest.

We carried on to the Douaumont Ossuary. This cemetery holds 130,000 unidentified remains of French and Germans brought in from the battlefield. The statistics are staggering, the crosses in perfect alignment only begin to show the wages of war. It's incredibly moving, and the building itself is quite spectacular. We climbed to the top of the bell tower and looked over the landscape because the memorial is right at the battlefield. My imagination just didn't work here at all. It's a hard place to consign it all to history. After something like that, we needed a break.

The museum is just down the road and after a coffee and sandwich, we left our helmets at the reception desk and immersed ourselves in the museum. The whole place is respectful of the sacrifice, the dead, the memory of the men and women of the Battle of Verdun. It's incredibly well done, and at your own pace, you can read, sit, contemplate and watch small movies in all languages. I think the enormity of the loss has a sobering effect on everyone who visits. We all wandered and whispered and took history in with every breath. Rather than interesting it was immersive.

Riding back to our hotel we had nothing to say. What can you say? I hadn't felt this empty since I visited the concentration camp, Auschwitz in Poland.

Our next day was going further afield. Boomie had read about a Fort and battleground that he wanted to see, so we went hunting for Fort Souville, then onto Vaux pool and the village of Vaux, or where the village once was because it had been obliterated off the face of the earth.

The fort was off the beaten track and had none of the entry fee, car park, or postcards for sale. In fact, is was a forgotten piece of history and falling down.

We scrambled inside and with our phone torch walked some of the alleyways until they were blocked with rubble. There was the feeling that the combatants had just left. On the outside, there was evidence of bombardments and trenches, dug-outs and huge holes from artillery shell scarred the earth. We could scramble all over the fort through the thicket of trees and it didn't take long to see the whole picture of shell fire. In 1916 it was all mud. Across the dirt road were trenches in the familiar zig-zag pattern. I believe if you dug just centimetres under the ground there would be bullet casings, iron and all manner of things.

We left to find Vaux pool. This was the site of the water supply for Vaux village and Boomie had an old book of Verdun that showed it decimated in the war. He wanted to see what it looked like today.

There is a memorial to the fallen at the pool, which is a big pond and it's peaceful. The village, a little further on, was rebuilt near the original, the old site is now a memorial.

From there we rode through the forest to Fort Vaux.

Fort Douaumont was the biggest and most prized, but our next stop at Fort Vaux, changed hands 16 times during the war and saw hand to hand combat. The French officer only surrendered when they ran out of water and food. As a mark of respect for the valiant fight, the German officer presented his sword to the French commander.

The lower levels of the Fort are flooded, the water still and clear, but the top-half houses a huge gun that would have been deafening at detonation. You can still see the marks on the walls as the fight inside took one corridor at a time.

Outside Boomie and I sat down on the grass to breathe. It's hard to take it all in.

I was ready for something different. I was ready for champagne.

Twenty three

I like champagne. It's not just the flavour, the fizz, but the whole experience of the tingle, the fragrance and the thought of drinking something that has a history behind it. I once won 6 bottles of Bollinger champagne in a writing competition and that educated me on cheap champagne versus expensive champagne. I'm not a snob, I'll get whatever comes my way in my price range, but if I had a million dollars … well.

We were heading into champagne territory and vineyards appeared to our left and right as we entered Reims. Our hotel was off the main highway which had to be negotiated at top speed and with a double, then a triple roundabout and our slip road a fancy manoeuvre, we were on high alert. Miss the slip road and the mistake necessitated a good 10 minutes before the next turn off.

The hotel was right next door to an 'all you can stuff down your neck' Chinese restaurant. We made a note to get back in time for the trough.

There is a lively city centre in Reims and the cathedral was put on the list of things to do. The Kings of France have been crowned at Reims Cathedral for more than 1,000 years, so I read. It is only 120 ish kilometres from Paris and in a beautiful part of the world, so you can see the attraction. The Romans liked the place, King Clovis liked the place and even Atilla the Hun thought it worth a look, although he put it to 'fire and sword'; probably saw the price of champagne and had second thoughts.

World War I did huge damage to the cathedral, as did the Second World War, and restoration work has not stopped since that time.

Reims saw the signing of the unconditional surrender of the Germans, (VE Day), on 7th May 1945. There was still bunting on the city streets from the anniversary celebration of VE Day a few days before.

There is a very large forecourt at the front of the cathedral and the wind whipping around in the morning was viciously cold. Groups of Chinese tourist huddled in the doorway ruining everyone's photographs. We scooted inside with a steady stream of visitors and readjusted ourselves out of the wind. We later found Reims was, because of its proximity to Paris and the champagne district, a drawcard; it was on every bus tour's itinerary.

We hadn't encountered so many tourists on our travels before and although it doesn't ruin the experience, it doesn't enhance it either.

Reims also has quite a few champagne houses, cellar tours and more tat than you can poke a stick at in the way of tea-towels, keyrings and souvenirs with the drinking theme. I was tempted, I'll admit to that, but we were on the bike, tat was something we didn't need or want. I had enough pamphlets to distribute through my underwear; I didn't need a tea-towel.

But ...

I found a wonderful pen & cigar shop, *La Régence*. A strange combination, but the man behind the counter was friendly as I 'licked the glass.' We got to talking and he told me the history of the shop, his love of pens and I decided to buy a pen from this nice man. It doesn't take much to convince me.

Jean gave me the royal treatment of sitting down in an alcove, trying pens on beautiful embossed paper with different inks, while Boomie wandered the shop marvelling at cigars that cost more than a pen.

I chose a Dupont, bright orange and loved every moment of the star treatment. We exchanged addresses, became Facebook friends, and I promised to send him a card from Australia next time I was there.

A pen book was thrown in for free, and I looked at my husband and the size of the book. Tricky on a motorbike.

'We can always post it.' Good thinking.

Time was getting on and we had gluttony on our minds. Not that we are gluttons, but we'd been living on sandwiches, three-minute meals and the like and a good Chinese usually comes with plenty of vegetables and if there is one thing we like, and we miss when travelling, it is our veggies.

I was looking forward to big helpings of broccoli, carrots and greens.

We were early and waited for the door to open, and about three minutes before opening the people began to arrive by the dozens.

'A party?' I asked a woman.

'Non, it is very new. We like.'

We'd only seen one other all you can eat at Le Mans. Apparently, it was a new experience for Reims. The place was huge and it was packed solid in about 15 minutes. You pay up front, then get a table and a drink and go for it.

It wasn't what we were expecting. It was Chinese food, you know, the western Chinese food, but it was mainly *hor'dourves.* Fried spring rolls, fried wonton dumplings, fried bits of fish in batter and fried everything that you could imagine. Fried brie cheese with a crumbed coating disappeared off the bain-marie like magic. Where were the vegetables?

I loaded up with noodles that had some greens in them and then we just did what the people of Reims did and went for cholesterol like we were snorting cocaine. If we ate up, we wouldn't need much tomorrow. I just squeezed in a dessert, which was typically French and beautiful to look at as well as tasty. Boomie squeezed in two helpings and we were done.

We staggered back to our hotel and flopped. Tomorrow we'd be going to Épernay. If Reims is the unofficial capital of the champagne district, Épernay is the jewel in the crown.

The road to Épernay is through vineyards, and a well-worn route. The town sits down in a small valley and as you turn the last corner on the brow of the hill, it lays before you like a watercolour painting. You descend with grapes either side of the road and the excitement builds as the names of the chateaux line the road.

There are a great many champagne houses to choose between for a tour. We went for a coffee and tried to decide which was value for money, the most on offer, open for business or had something worth seeing. Not an easy task.

Down Champagne Avenue, we peered into the immaculately kept courtyards of the wineries. Names like Mumm, Tattinger, Bollinger, Clicquot, Mercier and Pommery, Chandon and Moët whet the appetite.

A glass of these fine drinks isn't cheap. You can taste for 5€ for a 'small' glass, or get a 'big' glass for 10€ ish. We gawked at the people enjoying their tipple, and decided we should go on a tour with a glass included.

The house of Chandon et Moët looked as good as any, so we bought a ticket and waited with the others for the grand tour, then our glass of bubbles at the end.

The guide was a young woman with impeccable English and she treated us like we had money, even though we don't. There was none of the herding, follow me, pointing and don't fall behind. We were first told how to pronounce the name. Apparently it's, M-wet et Shan-do, as in d-o for orange. We tried it out, feeling superior. Now when I buy a bottle, I'll know what to say … not that I could afford a bottle, but it's nice to know you have the upper hand. It was all extremely civilized. We heard the history of the house, the family's connection to champagne and then went underground to the cellars.

Naturally, they were the biggest cellars, the longest corridors, the most bottles and I wouldn't have expected anything less. In the war years the village people took some of the bottles and hid them from the Germans, but with no lighting and 28km of tunnels once in - hard to get out.

At the end of our tour we were served two glasses of champagne, one white, one rose. It was good, but I'm not an expert and so the nuance was lost on me. We drank up and I bought a postcard, the only thing I could afford after the tour tickets. I was tempted, just a little, to buy a corkscrew as a reminder of our visit, but when the woman behind the counted said 20€ I erred on the side of caution and put it back. Tourist trails always see the customer coming.

The village of Épernay has a lot of charm, with fountains in squares, a boulangerie on every corner and a café on the other. The sun came out for our visit and we strolled about trying to look rich enough to fit in with the elite. We did see quite a few expensive cars, handbags and fingers with diamonds on them that would choke a horse. The in-crowd was definitely 'in'.

Back at our digs, we hadn't eaten, so decided to have another go at the trough next door.

'I'm going to pace myself,' Boomie said.

'Yeah, right.'

After visiting a laundry and getting our smalls folded by a woman, (she insisted it was part of the service) we were ready for 'International' travel. Our destination was Belgium and Menen.

Twenty four

It's all part of the Schengen area and so going to another country is as simple as driving on a road to your chosen destination. Back in the day when I hitch-hiked across Europe, I had a stamp at every border. Now, there is nothing to denote a crossing except signs in a different language. Just when you get used to something, everything changes.

In France, the signs tell you where you are headed noting the big town or city in big letters and all the small ones sometimes don't get a mention. The big town can be 200km away, but you are headed in the right direction on the right road and it's relatively easy.

In Belgium, the land of rules and regulations, the headquarters of the European Parliament and chocolate, the road signs tell you what the next village is. If you don't have it on your cheat sheet, you're up sh*t creek.

I'd marked the big names to look for, the exit numbers we needed, but couldn't see any rhyme or reason to the Belgium signs. All they told you was the next village, which as we were in the Flemish area had a lot of vowels in them. We were also on a highway heading to Lille which has a tortuous interchange of mammoth proportions. I wasn't looking forward to it anyway and now I didn't even know where we were.

'We should stop.'

Easier said than done when you are on a motorway and heading into a big city. There was no way off to the side and I really didn't want to take a slip road and end up somewhere else, like Luxemburg or Germany. Belgium is bordered by so many countries you could take your pick.

We slowed and waited for the next sign to appear, and it did, right behind a truck.

'Merde.'

I was beginning to think we'd never get off the ring road interchange when I saw something I recognised from the night before on the ipad, while making my cheat sheet. A large IKEA store loomed ahead and had Kortrijk written on the side. I couldn't pronounce it, but that was the name I was looking for.

We swung off the merry-go-round and ended in Kortrijk, any longer on the road and we'd be in Holland. Boomie pulled over and I revised my cheat sheet to include a few small villages. Now, I had some idea of where to go. We crossed the border three times circling around Kortrijk to find a way to the village of Menen. Eventually, we asked a local. She spoke perfect English, and pointed us 'just up the road.' We were so close to it we might have been able to spit on it, but as far as we could see there were no signs indicating we reached it. Perhaps people only go to Menen if they already live there?

Our stay in Menen was in a private residence. For our last place on our tour, I'd opted for something a little special. It wasn't that much more than a hotel, but it looked cottagey and super comfortable and a Belgium breakfast was included in the price. I'd heard they have chocolate for breakfast!

Menen is a little bit French, a little bit Flemish and a little bit Dutch. The signs are either one or the other two and the town is split. Although it is completely within Belgium it sits on the border and is so close to France people go to the shops in another country. Crazy, but true.

We found our accommodation and no-one was home, although I'd emailed and told her when we would be likely to arrive.

There was a note on the door, in Flemish and good ol' Google translate came to the rescue. We were to ring number six and they would have a key. They didn't have the key, knew nothing about it and were totally mystified about the whole process.

One good thing about Belgium is that most occupants, even the older ones, speak a bit of English.

I rang the number we'd been given and no answer. We read the note again and on closer inspection six was sixteen, the one a faint line.

Number 16 did have the key and the woman was the mother of the man who owned the building. She was lovely, like everyone's grandmother. She showed us the farmhouse within the walled garden and let us inside.

It was a complete mess with sardine tins and other fishy things over the table, crumbs and old coffee in the kitchen and it didn't take a great deal of deduction to see someone was already in residence.

Grandma was apologetic, profusely apologetic and rang her grandson to come and do something. We sat down to wait and the grandson duly arrived, apologising and explained we would be in the big house, they had Russians in the farmhouse. I've met quite a few Russians and they just love tinned fish. Don't ask me why.

Grandma made us a coffee while we waited for the lady of the house to arrive and told us the story of the farmhouse within the garden.

It was one of the first buildings in Menen. (here we go again, I thought). There was a small river running right by the yard and this was the toll house. Then, later it was a post office, and later a house for a large family. It was 400 years old and we were shown over the place, the little rooms, the nooks and crannies and the original staircase behind a fake cupboard to a loft. The grandson said it was the war.

'You know the war?'

'Yes.' We knew all about the war.

It was all charming. I'm a sucker for a walled garden, fruit trees trained on a brick wall and a pond. This had it all.

Boomie could see I was smitten.

'I know. A walled garden, a pond, the trees.' He knew me too well.

Our hostess arrived with more apologies. She had to take a friend's sick dog to the vet and got caught up in the drama.

The big house was beautiful and I pinched myself I wasn't dreaming. It had a bay window,

'I know, a bay window,' Boomie smiled.

It had a big ol' kitchen with a range and what looked like medieval tiles in the hall. The quilt was like a cloud, the bed deliciously soft and our room overlooked the garden. I think I was in heaven.

We hadn't eaten all day and so, after unpacking our panniers, went in search of food.

I'd been to Belgium before and knew they have a thing called a croquette. It's a filling of something, sometimes meat, wrapped in crumbs and fried. I could eat my own weight in croquettes.

Menen is small, but very multicultural. We found kebab shops, Chinese restaurants, Turkish flat-bread with an evil looking paste on it, but I couldn't find a croquette. We opted for Chinese and had a brilliant meal of vegetables. Lots and lots of vegetables with green tea and Belgium beer.

As everything was within walking distance we walked to a bar with the intention of trying more of the Belgium beer, and the Belgians like their beer. Some of it is very strong, some of it 10 and 12%.

The bar we walked into had three old men sitting in the corner drinking quietly and a barmaid with no front teeth. They watched us for a bit then plucked up the courage to ask where we were from. Australia was about as far as the moon to them. They couldn't quite understand how Australians might end up in Menen. The way it was signposted we wondered the same thing. We scooted over to their table and began to talk.

Dirk, Maurice and Freddy were great fun. As Freddy said, they had too many beers in their pocket, and so some of the conversation was pure seagull, but, none-the-less all great fun. We explained why we were in Menen, where we were going and what we hoped to see. They told us what to see, how to get there and were proud of their Flemish heritage. They spoke Dutch most of the time, Freddy knew English, but Dirk spoke Flemish to his mother and Maurice only Dutch. They lamented the way their culture was being eroded. I asked about croquettes and Dirk threw up his hands in despair. The last croquette shop had closed not too long ago and it was sadly missed. He went into great lengths on how to make a croquette, Freddy telling him he was wrong, the barmaid putting her two bits in on the method, Maurice at 85 y.o just shaking his head at the folly of youth and I'd started a war. It seemed croquettes were a sense of national pride, never mind there were several ways to make them. It was the same in Brittany with galettes. Everyone has a secret recipe or ingredient.

The men were the nicest blokes and after we'd had 'too many beers in our pocket', we parted company and promised to come back the next night for a farewell drink. It was a date.

Did I mention how fluffy our quilt was, and how soft the pillows? I never wanted to leave Menen.

But we did, the next day, to go to Ypres, now called Leper.
It is a good size town, but the two attractions, if that is the right description, are the Cloth Hall and the Menen Gate.
The Cloth Hall was bombed and shelled to bits in the war and has been lovingly rebuilt. The Menen Gate is the site where every night since 1928, a bugler has sounded the last post at 8 pm. It was stopped in WWII when the Germans occupied the area and moved, but since then it had happened every night at the gate in gratitude to the British and Commonwealth soldiers who gave their lives for the freedom of Belgians.
The square in the centre of Ypres is typically Dutch with the houses having flat fronts with fancy gables and windows. It's also typically Dutch in that it's clean. We were able to park in the square all day for about 2€, which is ridiculously cheap. I didn't need to walk about looking for a spot as there were designated parks for bikes. We found out that bikes are free so that was even better.

The Cloth Hall is a museum that has combined the kid thing with the adult experience. What I did like was the combined ticket that let us climb the clock tower and walk on the roof. The stairs are a killer and we were mouth breathing hard by the time we got to the top but once there, we could see the town, the countryside and get a view of just how flat the land is around those parts.
After all that exertion we went to a chocolate shop that was a feast for the eye as well as the stomach. It was

expensive, but we were in Belgium, land of chocolate and it would remiss of us not to indulge, just a little. And a little was all we could afford. You can pick the box size and then ask for one of those, one of that, one of the pink one and so on. We had about eight for our money, enough to say we'd tried it, without breaking the bank.

Walking the chocolate off was easy through the streets looking at the houses and the shops. I did find a stationary shop and bought a pen. It was on special. What can I say.

With about a two hours to wait until 8 pm we took the opportunity to go hunting for our dinner. We found a modern hamburger joint, all loud music and graffiti on the walls, but the hamburgers were brilliant. Huge things with everything in them and we could have split one between us if we'd know how big they were. With a beer and a burger the size of a dinnerplate I didn't think I would need to eat again for about a week.

A long time ago Boomie had purchased a print of the painting by Will Longstaff called 'Menen Gate at Midnight'. The original is a huge, haunting painting of soldiers as ghosts and their helmets as poppies. We'd had this picture hanging in our house before we moved onto the boat. The frames were fashioned from the timber of the HMAS Sydney by disabled servicemen after the First World War. It had some history and now we were standing right next to the Menen gate with a crowd of people waiting for the last post. It was a surreal moment I will never forget. There is a precision to the event. The crowd hush, the bugler steps forward flanked by two soldiers and he plays that mournful lament to the fallen. I looked around at the crowd and wondered did they feel the way I did? The whole thing was so moving, and after seeing all the battlefields, the graves, the museums, this was a fitting end to our trip.

We rode out of Ypres, sobered by war and pressed on home to Menen for our date with the boys at the bar.

They said they had a bet we wouldn't come, but we did.

We swapped addresses, talked over what we had seen, what we had missed and because we were riding the next day, decided not to get too many beers in our pocket. The barmaid let us out around 11 pm and we walked home; a perfect end to a perfect day.

Now we just had to go home across France in two days. I booked our stop in Rouen, just to make sure, and we headed west, right into two days of rain.

The only mishap was at a toll booth just after Rouen. I had kept the ticket in my glove and as I whipped my glove off the ticket flew out and there is no going back once you are in the gate lane. Boomie stopped and I hopped off and hunted for my ticket in the pouring rain as other cars swerved past me. I found it wet and flattened after being run down and ran back to the toll. You need to fit the ticket into the slot. Fat chance with a sodden ticket that was squished out of shape. I gently eased the ticket in the slot and crossed my fingers. Success. We pulled up on the other side of the boom gate and raced into the lee of a building as the rain came down. The only thing I could think of was our stove at home, warm and inviting. We had a long way to go and my underwear was wet.

We tumbled into Rouen, took our clothes off, cranked up the heater and steamed up the windows. Later I went out for take-away and we spent the night turning our clothes over on the heater for an early start. Our early start was with frost and ice. I will just say now, the makers of long underwear should all get a medal.

We hit the road around 7 am and hoped to be at our coffee stop in an hour. As we'd travelled this route before things were familiar and I knew where the best *pain aux raisin* were and where the coffee was hot.

By Caen we just wanted to get home. We'd been rained on for over 24 hours, sleet was sticking to us and the roads were filthy with muck which flicked up at every passing truck. It became an endurance run, but the nearer we got to Brittany, the better we felt. It really felt like we were coming back to our place, our home. Familiar signs and placenames began to appear on the signs, and I knew as the kilometres ticked by how far it was to our front door.

After the trip, it's never as cold, or as wet, or as bad, but I've never been so glad to get home to my bed, my pyjamas and my slippers. And the vegetables!

We did 2,700 kilometres in all, and as we sat down with a glass of red, and talked it over, it really was a wonderful experience. I made a note to send letters to our new friends with postcards I'd collected in Australia. But we didn't have time to sit back and relax. We were going to the Isle of Man in 10 days and needed to pack.

Twenty five

In ten days we had washed our clothes, stored the bike, packed our van and bought provisions for our trip to the Isle of Man.

It's a small island between Ireland and England with a long and convoluted history. It also has the wildest, most dangerous motorbike time trial race in the world. And that isn't a French boast.

People die every year at the TT race. It's the only race in the world that is true to the spirit of derring-do and sheer adrenalin rush danger. It is one of the races that motorbike enthusiasts flock to in their thousands every year. The little island comes alive with bikes, bikers, and all the merchandising mania you can wish for. We'd watched it for years on the internet. We'd read about it in magazines. I promised Boomie we would go on his 45th birthday. That never happened. Now, we were going, albeit 20 years late.

Our trip would take in Ireland, then across the Irish Sea to the IOM then over to the U.K. and home via Calais. A big circular route. We were camping with our little van at the

Rugby club in Douglas, which is the capital of the Isle. A van meant we could indulge in English bookshops and not worry about weight. Camping meant Boomie could park up and go out for a drink without always being the designated driver. This would be fun.

Booking in advance is a good idea if everything goes according to plan. You can guess, not everything went according to plan.

We were catching the ferry from Roscoff, a small village on the Northern coast of Brittany. Brittany Ferries were taking us to Cork in Ireland, but when we went to check in there was a sign telling us that the ferry had technical issues and wouldn't be leaving for 24 hours. Now, what were we supposed to do? The only good thing about the delay was that our booking in Ireland was still for the same day. Previously we were going overnight to Ireland, now we'd arrive on the same day as our B&B booking. No big deal.

After a telephone call I pushed our arrival time back, as we'd now be getting there at the end of the day instead of the beginning, and we went hunting for accommodation in Roscoff. So did all the other 400 passengers on the ferry. We jagged the last room in a hotel and went out to explore the village. It sits right on the coast and its claim to fame involves the young Mary, Queen of Scots, landing there on her way to Paris. You can see the actual steps, so they say, but I'm just a little sceptical about the claim. Still, it is a good story.

Roscoff being a seaside town, there are plenty of souvenir shops, seafood restaurants and bars. We walked the breakwater and then went for a meal of the local fish.

Our boat was leaving early the next day which stopped us from sampling the local wine.

Boarding the ferry is a long, slow, painful process. Anything to do with boats is a long, slow, painful process. The staff have it all down to a fine art, but the waiting is mind-numbing. When we were boating people we got accustomed to the wait. If you can see what the hold-up is, it's not too bad, but just to sit in a queue is bound to get you tetchy. Eventually, we were told to move through passport control, customs and our little van was put on board. It is a 14-hour trip. If we had gone in the evening the day before, we'd have done the crossing at night. Now we had all day on the ship.

Although I've lived on a boat, I still get seasick. All I need to do is get that first throw-up out of the way and I'm fine for the rest of the trip. I was dreading the ferry ride, no-one wants to throw up in public. So, I used a remedy that had worked once or twice before. You put an earplug in one ear and strap your big toe and the next together. Don't ask me how it works, but it does. It just might be the brain, psychosomatic, or some sort of reflexology thing, but it is my lifesaver. We had booked seats for the trip, thinking we'd be sleeping in them, and now in daylight, we really didn't need them; someone was in mine anyway, so we just walked around, sat in the sun out of the wind and diesel fumes and had a slap-up lunch from the restaurant. It was mostly English fare, everything with chips.

Coming into Cork it was cold and bleak, but the view was quite exciting. It had been years since I was in Ireland, Boomie hadn't been at all, so we were ready for adventure with a capital A.

Getting off the boat was just as mind-numbing, and once on land, it's not exactly clear where you need to go. We followed everyone else, always a good bet. If there is one place I hate to try and navigate it's a dock area. The GPS doesn't work, there are roads, and barriers, trucks and rail tracks and its confusing. It was a little more confusing because we were driving on the wrong side of the road for our van. Boomie is pretty good at that sort of thing, but it's still tricky. I kept reminding him to keep left, especially at roundabouts.

Cork is strung out along the coastal inlet and we just headed into town until something like a sign could point us in the right direction north. Our destination was just the other side of Tipperary. Just one left turn at exit 10 on the M8. Contrary to what the song might have you believe it's <u>not</u> a long way to Tipperary. 97km will probably do it. As the light faded we drove on and I put our B&B destination in the GPS. So far so good. It was around 10:30 when we finally arrived at the village and although our GPS said,
'You have arrived,' we hadn't. There was no house with a light on, no sign of a B&B anywhere.
We stopped at a pub, it was getting on to 11 pm, closing time and asked directions.
'Nev'r 'eard of it Luv.'
'Merde.'
I zoomed in on the GPS and we ended up in a housing estate. I knocked on a woman's door with the right number and she wasn't much help.
Our hostess wasn't answering her phone and now it was 11:30, we were tired and getting more than a little frustrated. I flagged down a man in a car and he didn't have a clue either. The case of the missing B&B. The driver did suggest that the hotel down the road might know. I had my doubts the B&B existed. They had our money and that was that. A scam, and the Ashwins were caught.
The hotel said they did have a room for 260€, the only one left. You've got to be kidding me. Another shoe salesman convention in a village the size of a postage stamp? It was after midnight and we gave up.

Boomie found a park near someone's front gate under a street light, we pulled the quilt from the back of the van and tried to sleep for a few hours. At 4:30 I gave up, the crick in my neck hurting too much and we decided to drive on to Limerick.

Limerick is a big city, but every hotel or B&B we tried was booked solid. It was weird. This was midweek, no holiday, no Saints day, not school holidays and yet we couldn't find a room. It was around 7 am and breakfast was calling, we'd had nothing the night before, except a few biscuits from our camping stock. Our love of Ireland was yet to kick in because we couldn't find anything open.

When I saw a petrol station Boomie pulled in and I was ready to eat jelly beans if I had to.

The bakery section was just starting and there was nothing on offer, but they did have fried things. We opted for a hash brown, cholesterol inducing sausage and chips and coffee.

The toilet was the stuff of nightmares. One was blocked with someone else's breakfast and the other locked. When ya got to go, ya got to go. We just had to leave our parting gift and,

'walk very quickly to the exit, do not look left or right, just walk.' I think Boomie did a three-point turn in under two seconds and we were gone in a cloud of smoke.

'I'm never coming back to Limerick,' Boomie said.

'There was a man in a van,
Who stopped to visit the can,
He left a big mess,
Of what you can guess,
And then shut the door and ran.'

I wasn't keen on coming back to Limerick either.

Nothing is very far in Ireland, and I'd booked our stay in Belfast, hoping to wing it for two nights from our scam B&B to the north. We stopped a dozen times along the way, hoping for a room, but there was nothing. Does everyone take their holidays at the same time in Ireland?

As we drove, the scenery was spectacular, which helped lighten our mood. We stopped at Loch Ennell and I made a cup of tea while we admired the view and tried to decide what to do. Our itinerary wasn't fixed at all, we could go any number of ways to Belfast. Boomie wanted to see the countryside, more than the motorway, and that meant cross country.

Around lunchtime, we found ourselves in Bellananagh, a small out of the way place, but miracle of miracles, they had a room at the B&B. The only snag, no parking. We would need to park our van at the rugby club and walk to our accommodation. That wasn't a problem in itself, but leaving the van overnight in an unlit car park was a bit of a worry. We parked up, locked up, deposited our bags and went exploring the town. On top of the hill was a large cathedral, well, large for such a small place, and we decided we'd take a look. There was a funeral taking place and we couldn't go inside, but that didn't stop a woman asking us if we'd like to come to the funeral.

'Ah, she was a lovely lass. Just 34 and died of the cancer. You'd be more than welcome. Come on in. It's standing room only.'

We declined graciously and headed back into town marvelling at the hospitality of the place. I've been invited to weddings when I didn't know the people, but never a funeral.

Across the road from our room was a tiny pub and we retired to have our first Guinness in Ireland. What better place to drink a Guinness?

The bar was one of those that was a converted front room, or parlour. It was small, had a bar and four stools and that was about it. Three of the stools were occupied with regulars so I sat and Boomie stood and inevitably the conversation got around to us.

'Australia eh?'
'Yes.'
'What do they do in Australia then?'
'Pretty much what they do everywhere.'
'Ah. So, they speak the English then.'
'Yep.'
'Ah.'
I think he needed to get out more.

I asked the woman behind the bar about the house and she said it had been a bar for many years and pointed to a small hatchway at the end of the room.

'That was the bar, down there. They served through the hatch.' The place was a rabbit warren of rooms, corridors and had an outside lav, so we were told.

It was just the sort of place we like to go for a drink, with a bit of history, a few locals willing to talk and nothing fancy.

'There's a ghost ya know.' The man who didn't get out much said.

'Really?'
'Ah.'
I waited for the story. That was all he was prepared to say. No-one else was going to elaborate.

We left them drinking and went back to our B&B to be met by the landlady,

'You'll be wanting breakfast?'
'Yes please.'
'A full breakfast?'
'Yes please.'

'A full Irish breakfast?'
'Yes please.'
'So, that's a full Irish breakfast.'
'Yes.'
'And you,' she inquired of Boomie.
'You'll be wanting breakfast?'
'Yes.'
'A full breakfast?'
'Yes.'
'A full Irish breakfast?'
'Yes.'
'So that's a full Irish breakfast.'
'Yes.'

I wondered what we'd get. I put a bet on fried food and plenty of it. I wasn't disappointed.

On the road again heading to Belfast we picked the largish town of Monaghan for our next stop.

We found a wonderful old hotel, again with a rabbit warren of corridors, but with a room vacant. They also did a great lunch in front of a roaring fire and of course, Guinness. We began to like Ireland. Our cheeks turned red as we roasted ourselves next to the fire and all was right with the world. After a second Guinness, we went for a walk and discovered what looked like an old wool factory powered by steam with those overhead wheels that take belts to run machines. It was derelict now, but by the size of it, probably employed most of the town. How things change.

After this stop we'd have two nights in Belfast, then the ferry to the Isle of Man.

Crossing the border from Ireland to Northern Ireland nothing much changes, except the signs are miles per hour instead of kilometres and they don't take Euros. They don't like English pounds either, preferring the Irish pound. We didn't have any cash and my French bank card refused to work in the ATMs although my Australian card had no problem, except for the problem that the card didn't have any money attached to it. I had Euros and the only thing to do was hunt out a bank and do some money changing for our Belfast stay.

'Ah, I'm not sure I can do that.' The woman looked at my Euros like I was offering pre-war Russian roubles.

'Do you have any pounds?' she asked me.

Sometimes you just wonder if these people should be let out in public. If I had pounds I wouldn't be asking for pounds. We decided we'd wait for the big city.

Belfast is congested, looks like every English big city with rows of terrace housing and really needs more roads, or fewer people. Our hotel for our stay was about 20 minutes from the city centre by bus and they assured me there was parking on the premises. We drove up and had to park a good two blocks away then walk as their parking was one tight spot in their paved front yard. Michael assured us the people would be leaving and the spot was ours for the duration of our stay which put a smile on our faces. The bus stop to the city was right outside the gate and the busses were every 15 minutes. Nice one.

We fell in love with Belfast. The main event in the city is the Harlan and Wolff shipyard where they made the Titanic. It's a Titanic overload, but interesting and they have quite a few free things to do. We saw all the free things, meandered over the dock area, but declined the museum which was full of school groups all racing around with question sheets and ice-blocks. The static, free displays were excellent with old photographs and not just of the Titanic, but how Belfast grew and how the population survived the daily grind. Walking back into the city I found a great second-hand book shop and indulged myself with a couple of books. Boomie bought several too, then we decided we'd have a cup of tea and a bun.

Who can resist a French patisserie? Not the Ashwins. The cakes didn't have prices on them, but we knew what they cost in France. It turned out what they cost at home was a pittance to what they cost in Belfast. The bill was enough to make you choke and we made a pact never to be fooled again; although they were very nice at the time.

Our last day was spent walking through the streets. We did want to see the cathedral, but with an entrance fee, it didn't feel the same, so we declined.

I'd been in Belfast in the 70s at the time of the troubles and seen soldiers patrolling the streets, barbed wire and checkpoints. I'd cycled around Northern Ireland with a girlfriend and missed a bombing by 5 minutes in Newry, nearly had our bicycles blown up by the British as we'd left them unattended and been taken to the bosom of Irish families more than once. All that seemed a world away. Now you can see where all that violence happened, just by catching a bus. We bought a round trip ticket and watched Belfast from a window, walked back to our room via a pub, met a couple who charmed the socks off us, swapped addresses and had fantastic fish and chips.

Tomorrow we'd be in England, albeit on an island, but we loved Belfast and fell a little bit in love with the people.

Twenty six

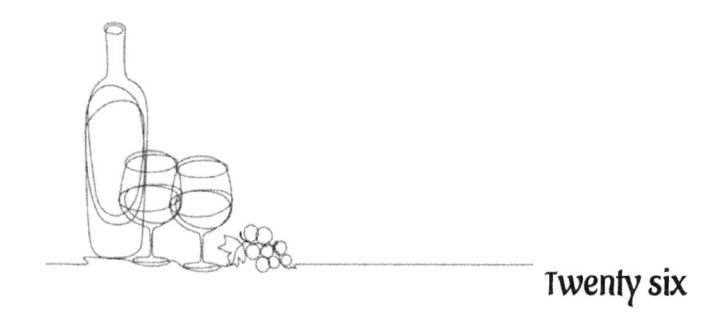

I put in my earplug, strapped my toes together and we joined hundreds of motorbikes in the queue to the ferry. We saw one of the riders of the race with his van and crew, and I had my first glimpse of the three-legged symbol of the Isle of Man.

We would be five days early for the race, tickets on the Ferry book out a year in advance and we were late booking. Some fans books from one year to the next.

It may be a small island, 30 miles long and 10 miles wide, but there was plenty to do, and once the race starts the roads close, so you are where you are for the duration.

The ferry lands at Douglas by swinging around the southern tip of the island and you get a good view of everything on the way into port. Boomie and I stood on the back deck and took it all in.

This event is the islands big income bonanza and they go all out with flags, bunting, and the theme of the motorbike race is over everything.

Tourists are a captive audience and so prices were on the steep side, but hey, we were on the IOM, a once in a lifetime trip. I don't think we could afford it twice.

I'd paid in advance for our camping and once again I didn't have a great choice on when to leave as when the race is over you can imagine the scramble to get off the island. We'd be there for three weeks. It was lucky they had a good bookshop and I purchased 'A brief history of the Isle of Man' so we would know what we were looking at - which helps.

Boomie and I picked a spot under a tree far enough away from the toilets and made ourselves at home. What I didn't realise was the tree had a sticky sap that dropped on our tent and van, making everything stick to it. Grass, bits of leaves, twigs, we were beginning to look like part of the landscape, but as the place filled, it was just too much trouble to move. The tree, a something lime, might just stop us having near neighbours.

From the Rugby club to the centre of town was a walk of about 2 miles uphill and as we had all day and nothing much to do we decided to walk it every day. At the very least we would get a little fitter and it would act as a counterbalance for the drink we might consume and the food we would probably eat, you know the sort, filth.

While waiting for the race days we toured the island to sample all its delights.

We could have driven over it in an afternoon, but decided to take the bus for the round trip. Busses usually wend their way through the small villages to get to the bigger ones. It's a nice way to see the place and not worry about parking, traffic and we could have a tipple at lunch. We went across the island and then back via Ramsey in the north. At

Ramsey, we had lunch in one of the hotels and it was a delight to see a group of old women come in for their pint and fish 'n' chip lunch. They had a hoot, gossiping, giggling and drinking and if I could be that happy when I'm eighty it would be an easy ride to ninety years old.

The towns on the Isle are full of character and quite charming. It was strange to see tough biker types sitting outside cafes drinking their lattes and eating cake, or buying tea-towels for their mums. You have an idea, by the look of the leather and tattoos that they'd sell their grandmother for a pint, but I think it's all just a front,' they are just mum's boys.

The thing a lot of bikers do is race the course before the real thing. The race is over the normal roads rather than a designated track, so it's easy to follow, and when you get up in the mountain there is no speed restriction for a mile or so. You can just imagine what sort of temptation that brings. There is a hospital on the island, and sometimes you hear a chopper overhead as another person has had an accident. News travels fast and the gossip tells the whole sorry story. Came off, concussed, broken leg, the bike a write-off, it all passes around the Rugby bar, then the campsite in quick time. We saw a few bikes loaded onto trailers, smashed beyond repair.

One morning, between the rain squalls, we drove the circuit, just to say we'd done it, the landmarks as familiar as our own streets back home because we'd watched the race so often.

The roads are very narrow and with Boomie on the off-side it was quite tricky not to get sideswiped by other cars and bikes. The bus has no fear and the driver pedals that thing around the lanes as if 3 inches from a wall is nothing to worry about. I flinched as he took a hair-pin bend and a truck was coming the other way. I could see the colour of the buttons on the truck drivers shirt and the crumbs on his collar from his morning tea. Someone behind me yelled out

'Shit.' We all released our breath and laughed off death.

Away from the crowds in Douglas the island has a quiet English charm we were told by a man on the bus. He had lived on the island for about twenty years and was now considered more or less a local. He said now the island sees a lot of week-enders with big money. They buy a house, but don't live there and therefore don't spend money there to support the local economy. It was a familiar tale. The same thing was happening in France. The English buy a holiday house that stands empty most of the year and a village dies.

Our unofficial guide was a Doctor and once he knew we were interested in history, told us he belonged to the local history association and they do archaeological digs, re-enactments and have lectures. Sadly it all shuts for TT week, but he gave us his telephone number should we come again. I promised to send a postcard once we were home. It is so much the people that you meet that colour your experience of a good time. So far we were having a good time.

In my book on the history of the IOM they talked about 'the thing.' The 'thing' is another name for a parliament or a coming together of people and the Islands parliament boasts to be the oldest continuous parliamentary body in the world. You can see where the thing meets, and has met for over a thousand years. It's not much to look at, but it's a 'thing'. Once a year there is a ceremony to carry on the tradition of the ordinary man petitioning his grievance to the thing. I don't know if he gets what he wants, but it sounds like a good idea anyway. Wouldn't it be great to call our parliament a thing?

'Get that thing outa here.'

On a blustery day we visited the southern tip of the Island and in that windswept place saw seals on the little Island called the Calf of Man and rabbits by the hundreds. My book informed me, people had been inhabiting this area for many, many years. The Island had been passed down, fought over, handed over, given to and taken from so many people, it's a wonder anyone knew who owned it. It was strategic, forgotten, strategic again and then abandoned by those who owned it. In WWII there was an internment camp for enemy aliens, but also a radar station listening post. For a little island, it had a very varied history.

What was equally intriguing was the belief in Fairies. All through the island there were fairy doors on houses, near bridges and waterways. The shops did a big trade in fairy dust, fairy magic and the like. Norse, Icelandic and those from the colder climes often believe in fairies and considering the IOM had a deep Norse history it wasn't surprising. But to see an old woman put out a small bit of breakfast for the fairies, that was surprising.

In the afternoon around 4 pm they shut the roads and it's full on practice time. We were at the Quarterbridge, a tight, almost 90-degree bend where the riders slow down. We soon learned you are either inside the track or outside for the duration of the practice. We decided to be on the inside so we could go home if it got too cold or there was a hold up. Hold ups happen all the time at races. A stone on the track or an idiot trying to cross, or a marshal who thinks he's Hitler and holds up the race ... because he can. That sort get short shrift from the crowd.

Our first glimpse of the racers just reminded us where we were - we were at the TT on the Isle of Man, FFS.

'What are the poor people doing?' Boomie asked as we waited for the next rider to brake hard at the bottom of Bray Hill and take the Quarterbridge turn on an angle of about 60-degree lean.

'Who cares!'

You can visit the pits which we did several times, but what we really wanted was a laundry. We found it and left our stuff with a lady who was quite disinterested in marking it as the Ashwins. I briefly wondered if we'd ever see it again. When we did collect it, they had ironed our underwear, there is a first time for everything. Boomie was most impressed.

'Don't get used to the idea,' I said.

We also became merchandise sluts again. You really can't help it when you know this is the one and only time you will be there. We bought t-shirts, hats and then another t-shirt and another hat. We had the van, so weight wasn't an issue this time. I loaded up with post-cards, free postcards and naturally, free pamphlets. (would you really expect anything less of me?)

The weather turned nasty two days before the race and there is nothing charming, nice or exotic about camping in the rain. For with the rain comes mud. In a camping ground with hundreds of people all tramping to the clubhouse and toilets it soon became a quagmire ... great for skidding on with motorbikes and doing mud burnouts, but terrible if you want to go to the showers and try and keep your feet clean, or anything else. Everything turns to mud as the rain came down and didn't stop.

They said the race would stop, even dare-devils don't chance the rain on a road that has the normal detritus of oil and other stuff.

We took the opportunity to visit a motor museum in Jurby on the northern end of the island.

We were first in and spent the whole morning looking at motorbikes, cars, and engines. It's a private collection and there was so much interesting stuff to see and read, we went out, had a coffee and muffin and went back in again. They had big American cars like the ones from Ghostbusters, and cars from gangster funerals aka Al Capone. They had things like the last car off the production line and prototypes of other cars that were wildly futuristic. We agreed it was the best motor museum we had ever seen. I usually get bored after a bit, 'cause you can see just so many overhead cams and shims, but this museum had the human story to go with the car. We met the father of the owner prowling around in his wheelchair, and he was a delight to get to know. He let us inside a Greyhound bus that was usually off-limits, telling us how he went to Nevada to get it. He showed us a car that his son had wanted for ages, but the old lady wasn't for selling. Eventually after about 7 years he got a phone call and she said if you can come now, you can have it. He was on a ferry before she'd hung up.

It is those types of stories I enjoy. The human touch.

As the rain cancelled the first day of the race we decided to go to Peel on the west coast for the day. It was here we found a small shop and got to talking to the owner, Allan. He was a wealth of island information and had memorabilia from the TT going way back, postcards and coins from the 1800s and original paintings from local 'famous' artists. Next door to him was the most fascinating second-hand shop we'd encountered in a long time. I found an antique fountain pen which was a steal and a brooch that the man gave me because I bought the pen. It must be my honest face. People do nice things all the time.

We spent the rest of the morning rummaging through that shop and could have bought so much we didn't need, but wanted. I saw books with wonderful inscriptions, a typewriter that looked ancient, war medals, hat pins, funeral wreaths made of human hair, sheet music from the 1800s, china dolls. Boomie saw weird and wonderful tools, some for pulling teeth, a motorbike helmet that was worn by a famous racer and we did umm and ahh about a stuffed mongoose fighting a cobra snake. What a conversation piece!

There is a castle at Peel which we walked around and then went for fish and chips down by the sea watching the fishermen come into the lock, yachts go out to sea, and telling ourselves how lucky we were not to own a boat any more. It was the first time we'd talked about how we felt about not owning a boat. Sad as it was, we were both glad that chapter was done. Boating was expensive, we were always anxious for the next mooring, the next marina and always, always watching the weather.
Now, all we had to do was pity the poor soul who thought it would be fun to own a boat. Buy in haste, repent at leisure.

The race was beginning to be a washout as the weather didn't improve. Some people were considering leaving early before the race proper had begun, others just packed up and were off. We all watched the weather forecast for a break.

The weekend of the race the population of Douglas swells in the thousands. If you want a spot on the track to watch you needed to be early. We thought we'd hike to the next bend and had seen the small grandstand there was only £5.00. That seemed do-able.

'Oh,' I stood there with my mouth open. The £5 was for practice. The race price was three times that for a bit of wooden plank. Still, this was the TT and we were on the Isle of Man, FFS. We paid and sat with a good view of the S bend at Braddan. The church ladies did a nice cup of tea and sausage in bread and there were toilets, always a good thing. The weather cleared for one race and then there was a fatality and the other races were stopped until the crash was cleared away, then it began to rain. They managed one more lap and it was all over for the day. I listened to the talk of the people around us, but the mood was all doom and gloom. Weather - from the South, from the North, from the sky. It was predicted rain for the next week and probably the week after that. But, this was the TT and we were on the Isle of Man FFS, we would enjoy ourselves regardless of the weather.

At the end of the racing day, the boy racers wait for the road to open and as soon as it does they follow the real racers around the circuit. They take off like they are the professionals, and some of them are very good indeed. They have the whole track to themselves before the cars begin to trickle in and it must be thrilling. If I was younger I'd have a go myself. Well, maybe not!

Boomie and I once did a lap of the Philip Island track in Victoria, Australia, but that was at a much more sedate pace.

In the week leading up to the race on Saturday, they had only managed to have about two time trials. The weather was shocking for racing. There was a rumour going around that the whole thing would be cancelled, but I couldn't see that happening, there was too much money at stake. We'd been on the island for two weeks and only seen a couple of races.

The time trial at 4 pm was cancelled and it was looking like they might not have another race and still the rain came down.

We sat in the tent and tried to rationalise our options. The race for tomorrow was postponed, the sidecars cancelled, the time trial cancelled. It was a wash-out. Some people were packing up, the mud, the rain, the lack of racing all too much. We had another week and a bit to go. We were spending money for the hell of it and we'd seen all there was to see.

I opened a bottle of red and we watched the camping ground thin at an alarming rate. People were throwing their tents into vans, on bikes and going home.

We decided to join them and pack it in.

If we could change our tickets we might be able to get off the island. I didn't hold out much hope, knowing how hard it was to get a spot off the Island in the first instance, but we'd go and ask. If we couldn't get a ticket, so be it, if we could, we'd take it.

The woman at the Steam packet company said we would be on a standby basis. We put our names on the list and rushed back to pack up the tent in the rain.

'Throw it in, we can sort it later.' And that's what we did, rushing back to the ferry terminal and waiting with other hopefuls.

It's like waiting for your bingo number to come up. All eyes are on the man who walks through the door with cancellations and spare spots. The 5 motor bikers waiting had more chance than us, we had a van which needs more room. Their number was called and they whooped and ran to the ferry. We waited, looking at the fellow with the clipboard.

'Ashwin,'

Boomie and I ran to the van and we were the last vehicle to board, our van right at the very edge of the boat near the roll-on, roll-off door.

On the boat I received an email from the shonky B&B. Refund in full. That made me feel a little better. Apparently we were, 'right there'. Knowing the Irish, I wasn't sure 'right there' actually meant 'right there'. Still, we had our money back, a nice surprise.

We were going to Liverpool. Luckily I hadn't booked any accommodation for this part of the trip. There is usually a pub or B&B in a village in England, well that had been my experience 30 years ago anyway. Hopefully, some things never change.

Twenty seven

Did I mention how I dislike ports? Liverpool dock is horrible. There are no discernible roads, and unless you know where you are going, way too many options. We just followed the bikes in front until we came to a proper road then under the Mersey River to end up at a toll booth - for busses. Nothing worked and we couldn't understand why. A bus came up behind us and now we couldn't reverse. The woman on the intercom was wah wah and the bus driver leaned out of his window and said,

'Bus lane luv,' and pointed to the sign. Of course, if you were in a bus, you could see it, way up there. In a little van, it wasn't that obvious. Well, not to us anyway.

I've never been so glad of our French number plates. I acted French, doing the throwing up of my hands and apologized in French. The bus driver I'm sure called me a frog and backed up. Boomie did a fancy manoeuvre and we tried again in the proper toll lane for cars, God only knows where we were headed now. Then by luck rather than design, we were on a motorway, out of town, heading south.

When I'd been planning I had a vague idea we'd go south via Wales and end up somewhere around Lyme Regis. I was keen on seeing the fossil beaches, having read the story about Mary Anning and her discovery of the ichthyosaur. From Dorset, we could follow the coast to Dover, about 220km away.

Our first stop was Chester or thereabouts, hoping for accommodation.

You'd think I would have learned my lesson by now, but we couldn't find anything within our price range in Chester. Oh, we could stay at £150 a night at a dozen places, but we were more budget-conscious that that. Every B&B we tried was booked. All the cheap hotels were full. I guess everyone has the same idea of budget-friendly accommodation. We might have camped, but to set the tent up for one night after a long day, wasn't my idea of fun, nor Boomie's.

Perhaps there would be something on the way to Wrexham the next big town. We stopped at just about every B&B along the way looking for something, but that shoe salesman convention looked like it stretched all the way to London.

The saving grace was the long twilights in the northern hemisphere. The countryside was quite spectacular in the low light and the names of the villages right out of every P.G. Wodehouse book I'd ever read. I mean, who wouldn't want to live in Cockshutt or Picklescott. We ended up in Church Stretton, a small village that was quintessential English and went to the pub for some advice, and maybe a room. They were refurbishing the rooms, just our luck, but the landlady pointed us to a B&B across the main highway. The B&B was full, but the lady very kindly pointed us to another village that might have accommodation.

At Upper Affcot I asked in a pub which advertised rooms. Nothing, but he did know a woman down the road. This was getting ridiculous. He rang and she had a room.

'The name is Ashwin, and don't give it to anyone else, we are on our way.'

With a mud map written on a napkin because he didn't know the name of the place, or what street it was on, but knew the woman whose husband played bowls with his wife on a Wednesday ... and so it went on. We just wanted a room and a bed, not his life story. Boomie was circling the block because the small lanes were too narrow to park and I really didn't have time to listen to a blow by blow account of the tight finish of a game of lawn bowls. We followed the directions, only going wrong once and then found the house next to the pub with the red roof and those flowers out the front that smell and are pink. This must be it then.

Our B&B stood next to the pub with the smelly flowers, whose claim to fame was attached to Lord Nelson and how he may or may not have stopped there on the way to somewhere else. It was flimsy at best, but like the French, you need to take it when you can.

Waking up to a cooked English breakfast after a wonderful night's sleep on a dreamy soft bed is the best way to start the day. We were the only guests, the shoe salesmen didn't get this far, so we had the breakfast room to ourselves and could talk to our host. The house was around 200 years old and was a staging post, post office and pub, before the one next door took that honour. It's always fascinating getting a little back story with your breakfast. The room was decorated with prints and photographs of soldiers, all of whom were family. We left with a cheery wave and a bright sunny day was on the horizon. I like small village life. English small village life has that feeling of permanence and history attached to it, and they speak English, always a bonus.

Our final destination was Dover so we carried on, heading south through villages, small towns and eventually hit the motorway. Dover is signposted from way out and that makes things very easy. Being able to read the signs even better.

Gloucester and a cup of tea put us in the right mood for the motorway heading to Reading. So much for seeing some of Wales and Dorset. Next time maybe.

All the familiar names whizzed by, and by the time we were on the outer ring road of London, we thought we'd just carry on to the coast, maybe stay a day or two around Dover and Folkstone, have a break from driving and sample the delights.

We arrived in Dover at 11:15 and went to a travel agent to book a ticket on a Ferry. It was leaving in thirty minutes, the ticket was a half-price stand-by, and the dock was just two minutes down the road.

Sometimes it happens like that. Who can pass up half-price? I put in my earplug, strapped my toes and we joined the queue for the ferry to Calais.

Once onboard Boomie and I treated ourselves to fish and chips watching the white cliffs of Dover disappear. It's such a short distance that in the middle of the channel you can see both sides at the same time. The maritime traffic was everywhere in that short crossing and we briefly mused on sailing the channel ... only briefly.

Another port, another lot of roads that make no sense, but this time we were funnelled so effortlessly onto our motorway I wondered why I was anxious at all. The A16 shoots down south and it really is that simple to just keep going. Knowing that the roadhouses would be well signposted, how the tolls work and once again being on the right side of the road for our steering wheel all felt very comforting.

We decided to knock off the trip in one go. Calais to our village was just a smidgen over 600km. On a motorway, at 130km/h the distance just flies by.

'You know we really should be on the Isle of Man.'

'I know.' Somehow our escape felt deliciously naughty. We'd heard that they only manage one race on the day. That made us feel a little better at our decision to quit while we still had some money left in the bank account.

Our GPS said we'd be home in time for tea and we were already thinking of vegetables, lots of them. Travel plays havoc with your digestive tract.

Twenty eight

It was time to get stuck into our renovations again, hopefully, so we could have the whole summer to see things.

We tiled our toilet and bathroom, then blitzed the painting and lambris. The old house was looking very smart and our neighbours were impressed. We had put the Australian touch to the colours, the curtains and the light fittings which Myriam thought were innovative and different.

I had invited our back neighbours for a look one afternoon and Marie-France loved what we'd done with the old bathroom. She brought her husband to have a look and they invited us over for drinks that evening.

As I mentioned, you don't usually take anything when you are asked for drinks. Flowers or a pot plant are ok, but Marie-France had the most spectacular garden, so giving her a plant felt a bit silly. It felt strange not to bring something, so I made a batch of little cakes and put a ribbon on a box, sort of a fancy present and when we arrived I presented it.

'Thank you.' Marie-France only knows two words in English.

She showed me the dictionary on the table and I pulled out my phone and punched in Google translate, although after a few drinks we were usually all on the same page. Alcohol can make anyone bilingual.

We were served a home-made drink in a tiny glass. It was a brandy that was infused with berries and herbs and one small, tiny glass was never going to be enough, but that was all we were offered. No wonder no-one gets fat. We ate cute little pieces of cheese on tiny little bits of square bread and little round biscuits the size of your thumbnail. It was very civilized, very chic and very French. In Australia, there are lashings of dips, cheese, celery sticks, and biscuits. That's not to say everything wasn't wonderfully delicious, Marie-France is an excellent cook, it just didn't bode well for a convivial evening.

When all the snacks were cleared away it was a red flag that the evening was at an end.

'We gotta have them over for our version of drinks,' I said to Boomie.

'I don't think they'd like it.'

'We can but try.'

With things going so well, it was time to treat ourselves to a lunch out.

Our friends had taken us to a small restaurant a year or so ago and it was this we would try to find again. I could picture the place, we just didn't know where it was, but that never stopped the Ashwins.

We never did find it again, but we found a wonderful small restaurant at Mer-de-Bretagne that was to be our new favourite. It was situated opposite the church with a view of the square and the food was to die for. I had galettes and Boomie a Caesar salad, but it was the dessert that really impressed. We had a banana split each, the size of a dinner plate. It had three flavours of ice-cream, wafers, bananas, sprinkles, nuts, drizzles of chocolate and chocolate chips, and some sort of liqueur. Then over the whole lot was a sparkler and paper umbrella.

'Did we order this?' Boomie looked at the towering delight.

'I guess we did.'

It was a struggle, but the Ashwin's were up for the challenge. I expected a round of applause as I put the last bit of chocolate in my mouth.

'We need to walk.'

I could hardly move.

Everything was closed for lunch in the small village, so we retired to the church grounds and sat down in the sun until I could breathe without letting out my trousers. From little squares that might not feed a sparrow to a dessert as big as my head. Something in-between would have been nice.

What we did have 'in-between', was a common drain. On one side we sat, on the other Myriam.

It was early morning when Myriam knocked on our door for help. Her toilet was blocked.

STAND BACK. The Ashwin's knew all about blocked drains.

She had been struggling with the toilet drain for a few days and now it was completely blocked. They were going outside and using a plastic bag and a bucket.

We all stood on the footpath and peered down the manhole. Boomie fetched our hose and we fed it into the hole, but things were bunged up pretty bad.

We used our water-blaster to get a good head of water, but it just wasn't working. We'd only dislodged a little bit, when we wanted the motherlode. Boomie tried our flexible rods and there was a great whooshing, which whooshed over all the audience. Of course, Giullaume thought it hilarious.

But the pipe was unblocked, we were heroes and our expertise in all matters of the toilet supreme.

We were treated to an evening of drinks at Myriam's by way of thanks and I didn't leave anything to chance, and came prepared with wine, snacks, more snacks for her boys and she'd made those little tiny squares with little tiny bits of cheese on them. I wondered if everything was kept small because they were afraid of the toilet repercussions should they over-indulge. Our drains were A1, not a problem in the Ashwin house.

Our expertise spread far and wide and when Nadia's toilet was blocked, who ya gonna call?

Nadia, Nassim and Yasmin lived three doors down and couldn't afford a plumber. The man of the house had chronic diabetes and was in dialysis three days a week, Nadia had to do everything, but her blocked toilet had her beat.

Myriam called us and we came over to take a look, being the experts in such matters.

The toilet drained like ours, into a sump open to the air, then cascaded into a pipe and that was as far as it went. There was a large mess. We'd seen it all before, but Nadia was mortified. I knew the feeling as half the street had seen our poo problems a year or two previously.

The pipe blockage wouldn't yield to our prodding and poking and so the only course of action was to dig. Myriam went for a piddling little spade that was more suited to digging a sandpit.

STAND BACK. The Ashwin's have tools.

Our spade was a long-handled thing, and together with our pick and a shovel, we began to dig.

You would think that if you just follow the angle of the pipe you'd soon hit it, then just follow it.

We knew there must be a box holding up things. There is always a box.

'Non. A boîte?'
'Oui.'
'Non.'
'Oui.'

We started to dig looking for the dreaded box, Myriam and Nadia not believing us. The hole was getting bigger and I began to doubt the logic of a box when we saw the pipe. Now, there were two pipes going off in different directions.

'That's French plumbing for you,' Boomie said.

The hole got bigger as Nadia's front yard disappeared under a mound of dirt and we were getting deeper.

'It's gotta be here,' Boomie said. I took another bearing from the open drain.

'I reckon it's over there more.'

The hole got bigger.

We put all our logic to the problem, but French plumbing doesn't follow logic, probably never would.

We had a decent size hole in the yard and then,

'Boîte.' Myriam pointed.

We all felt it was only right that Nadia should take off the lid, it was her box after all. She bowed out to Boomie.

It was like some sort of archaeological discovery as we all looked down the hole at the offending box. The pipes from the house came down to the collection box. Once the box was full it would overflow down to the street in the pipe.

'How do you know the box?' Myriam asked. I gave a Gallic shrug.

The collection box was full of roots and poo. Nadia died a thousand deaths of embarrassments. I knew that feeling oh so well. We cleaned it out and instead of burying it we put a big rock on it so in future she would be able to get to it. The kids were fascinated with our efforts, but the toilet was fixed.

We were invited at a later date for afternoon tea. It was a date.

Nadia's afternoon tea involved Algerian snacks. Delicious things made with flour and cardamom seeds and then some special Algerian biscuits. We also scored a bottle of wine in thanks. As we were enjoying our biscuits the mayor of the village stopped by and although no-one spoke English and we only had the most minimum French a great time was had by all. Who needs 100 hours of lessons when all you need is biscuits, wine and good company … and a blocked toilet.

It was when I was toting all our empties to the municipal recycling bin that I thought our wine consumption was getting out of hand and so we made a pact to only drink on a Friday. France has a big drinking problem we heard from the local doctor.

With our new resolution in mind we went shopping for our Friday bottle at the supermarket. I usually wander on my own when shopping. I know what I want, where to get it and what brand I like. Boomie goes off in search of 'things'. This particular day he came back with wine and we went to the checkout as usual. When the bill was presented it was quite large, and I couldn't understand why. We had wine - 3€, meat - 5€ and other sundries. I paid and looked at the docket. Boomie, without his glasses had chosen a 49€ bottle of wine. A costly mistake, but at that price we were expecting top quality and, well … it didn't happen every day. The wine was no better or worse than our usual 3€ bottle after all the build-up of the price. Not that we know much about wine, but we know what we like. I made sure Boomie took his glasses next time. He reminded me about the animal meat. It seemed we were even.

And it was while we were drinking our expensive wine, thinking over all the good things that had happened, all the people we'd met, we got around to something that had been niggling at us, bothering us for a while.

Boomie was coming up to his aged pension age, we weren't getting any younger and the Australian Tax office might want some of our money once we told them we were French residents. We also had an aging parent, and a family member that needed support. It was a big call. We cracked another bottle of red and thought about our options.

Why is that just when you think everything is tickity-boo someone puts a spanner in the works.

We were also worried about the situation regarding foreign nationals in France. Brexit had stressed the system, and there was talk of a price hike for visas, and stricter controls. We were safe for now, but how long that would last would be anyone's guess.

Health care in France is a system whereby you are covered by the state for a minimum amount, but you purchase insurance for the gap. Whether you purchase to cover the whole gap 100% or just a portion is up to you and you pay the out of pocket bit. We had private health, a visa requirement, via a company that had doubled our premiums in one year. Our money was being pulled more than one way and if things carried on, all the things we came to France to do would be hard with less money, or near impossible.

Also, France as a country was broke. Prices were rising in just the four years we'd been there, the yellow vesters were protesting, and for the ordinary citizen things were getting tough. The European Union was making things difficult too with more visa red tape and requirements.

You can guess where all this is heading.

We had a big think over a couple of weeks. We went to see a chateau, took a ride on the motorbike along the Nantes-Brest canal, went to Quimper for a long weekend and at the end of July we'd decided to go back to Australia.

That momentous decision necessitated selling everything. Lock, stock and barrel.

Twenty nine

Of course, there were a lot of things that we didn't want to sell, but we knew with the bio-security in Australia anything organic might not make it through quarantine. I started the job of ringing around for quotes. Once I knew what we could take, or more to the point how much we could afford to take, then we'd know what we needed to sell.

There is a site in Brittany on Facebook for English speaking bargain hunters. The buy, swap, sell formula works all over the world. The is also a site called Leboncoin, a sort of French wide E-Bay buy, swap, sell.

I tried various removal people and half didn't ring back, the other half thought we were millionaires and wanted to go the moon. When we had shifted our gear to France from Australia, door to door was relatively cheap. Now it was relatively expensive. There would be some serious culling needed.

We put a sheet on the shed floor, measured it and put everything we wanted to take within its measurements to a metre high. With an accurate metre cubed we would get a decent quote. Our quote squeezed into our budget with not much left over. I began to get creative and put things in things, because it's not so much the weight as the volume. It's amazing how much you can fit in a shoe, a saucepan or a camping box.

We'd bought second-hand furniture for our house and now I advertised it for the same price I bought it. It all disappeared quite quickly. One woman was just setting up house and bought as much as would fit in her van. We delivered the lawnmower the next day.

It was hard seeing all our stuff go, when it felt like we only just bought it, although in reality we'd had it four and a bit years.

Myriam was particularly sad at our decision. We were, she said, the best neighbours she'd ever had.

We naturally had to sell our little van and our motorbike. They were the hardest of all to let go. We'd had such fun riding and driving.

The man who bought the bike was so happy. He kept grinning to his wife and exclaiming,

'Impeccable. Impeccable.' We knew he would have fun on that bike.

The young lad who purchased our van was smiling so hard his face was frozen. He couldn't believe his luck to get an almost new van.

But the thing we needed to do was sell our house.

I thought buying a house was hard. Try selling one in France!

To sell a property you need a real estate agent. Well, you don't need one if you know all the rules and regulations. We didn't and so, an agent was a necessity. As luck would have it a new agent had opened up in the village and he was English, his wife French. The best of both worlds.

The paperwork is every bit as onerous as when you buy a house, only this time you need to prove not only who you are, but the credentials of the house. Is it the house that they say it is on the documents? Is it the house that has been described as such or is it different?

And this time as we were selling, not buying, we didn't pay the advertising.

I knew we had a good house for the market. When we'd travelled around looking for a house, we saw some horrors. Our house would be presented in its best possible light, a thing the French don't seem to do. It's almost as if they don't want to sell their property. Dishes in the sink, unmade beds, grass overgrown.

Our little house would be polished. It was newly painted, new curtains, new kitchen, bathroom and toilet. There would be fresh flowers, the garden neat and, well, we loved it, someone else surely would.

We had to engage a solicitor (*notaire*) to do all the conveyancing on our behalf and sign documents on our behalf if the house was still for sale when we left.

They have a system of Capital Gains Tax in France and the agent related horror stories about people paying thousands of Euros. There was a sliding scale over the number of years you had owned the property, but we were top of the list. The offset is the renovations. I had kept all the dockets from our purchases and so I put them in a spreadsheet, photocopied all the receipts and sent them along to the Notaire for they are the ones that disburse the funds once all the bills are paid.

There was the matter of the technical surveys to be done on the house. Electricals, asbestos, drains and foundations. An appointment was made and a bright young fellow came along and took samples, went into the roof space, dug a bit of dirt at the foundations and looked at our new electrics. We paid his bill hoping for no problems.

Before all our money ran out, I booked our tickets to Australia. Now it was really final.

Finishing up the small bits and pieces on the house for someone else just didn't feel the same, but we knew it's the small details that make a difference.

The thing that gave us comfort was when the agent brought his wife through he said,

'You are going to be surprised.' She, being French, was expecting a dingy little old house. What she saw was a modern, colourful, functional house. Her eyes widened and she smiled. She especially liked our light fittings and our newly painted front door. I couldn't see anything remarkable in either, but then I'm not French.

Now we just had to wait to snag a buyer.

And while we waited we saw a house on the internet in South Australia. It ticked all our boxes, especially the price.

So we negotiated and bought it. If you say it quickly it doesn't seem too hard. There was a mountain of paperwork because we were in France and had to have all our documents from the French bank officially translated. We had to have the Notaire declare we were who we said we were, and sign a stat dec and get it translated. But with email, you can do anything. Like, buy a house over the internet.

Then our goods left the building. Now we were living out of our suitcases until our departure, the only small light at the end of the tunnel, our washing machine was being sold with the house, so we could still do our smalls.

Every week we checked in with the agent. Nothing. Every week they asked for a document, or file. I had it all, I'd kept everything and could whip it out at the slightest suggestion.

They asked for the details of the previous owners. I could tell them when they were born, when they died, where they were buried and what was their favourite rum. The agents were impressed.

'Give me another, go on, I'm ready.' I had our whole life history on file, and I assumed everyone in France kept a filing cabinet in their house. The French love their paperwork.

It was a week before we were due to leave, someone put an offer on the table. We took it.

It takes quite a while in France for all the documents to go back and forth between the parties. The prospective owners wanted to see the roof space. They loved our kitchen. They saw our light fittings and gushed. I'm convinced our light fittings sold the house. We'd got them cheap and they were just modern, round light fittings, but what do I know, I'm not French.

As our days ticked off, we cracked the 2010 bottle of wine Daniel had given us.

'À ta santé.'

It was delicious.

The Notaire would finalise all the signing.

I'd booked one night in Paris before our flight, just to take home some lasting memories.

This was the end of one adventure and the beginning of another.

The Ashwins were scheming again.